TRANSFORMING IDEAS

TRANSFORMING IDEAS

Selected Profiles in University of Rochester Research and Scholarship

Edited by

ROBERT KRAUS

and

CHARLES E. PHELPS

UNIVERSITY OF ROCHESTER PRESS

First published 2000
by the University of Rochester Press

The University of Rochester Press is an imprint of Boydell & Brewer, Inc.
668 Mt. Hope Avenue, Rochester, NY 14620, USA
and of Boydell & Brewer, Ltd.
P.O. Box 9, Woodbridge, Suffolk 1P12 3DF, UK

hard cover: ISBN 1–58046–058–5
paperback: ISBN 1–58046–090–9

Library of Congress Cataloging-in-Publication Data
Transforming ideas : selected profiles in University of Rochester research and scholarship / edited by Robert Kraus, Charles E. Phelps.
    p. cm.
    ISBN 1–58046–058–5 (alk. paper): ISBN 1–58046–090–9 (pbk.):
    1. University of Rochester—Anniversaries, etc.   2. University of Rochester—Research.   I. Kraus, Robert, 1951–   II. Phelps, Charles E.

LD4721.R549.T73 2000
378.747′89—dc21   00-030256

British Library Cataloguing-in-Publication Data
A catalogue record for this book is available from the British Library

Designed and typeset by ISIS-1 Corporation
Printed in the United States of America
This publication is printed on acid-free paper

Contents

Preface

Perhaps nothing in the long-run progress of civilization stands as more important than the development of new knowledge. At the very center of that progress, since the dawn of modern civilization, have been the universities of the world. In the United States, the emergence a century ago of the modern research university brought a major shift in this process. While public universities grew through Land Grant and other mechanisms, private higher education arose in parallel as a vital intellectual force. Perhaps most important, and uniquely in the United States, higher education developed in a competitive environment (compared with that of the state-owned and state-supported systems throughout the rest of the world). Competition both for the best students and for the best faculties in universities and colleges across the United States soon swept our higher educational system into a position of world leadership. Indeed, as a success story in international competition, no sector of the U.S. economy comes close to matching the dominance of U.S. higher education in the world (save, perhaps, professional basketball).

In the midst of this dramatic transformation of higher education in the United States, the University of Rochester began to emerge as a second-generation participant, forcefully launching itself into the world of a modern research university in the post-Korean War era, under the leadership of President W. Allen Wallis. By that time, the University was already renown for its strong medical center and the world-class Eastman School of Music (both created during the presidency of Rush Rhees in the 1920s, with the transforming gift support of George Eastman). Eastman's residuary estate gift and other major gifts (including those, for example, of Joseph C. Wilson) subsequently created the resources for transforming the institution into an active and dynamic participant in the world of research universities.

With the growth of the intellectual force of American higher education came an improved understanding of the way in which knowledge increases. The classic "scientific method" story we all learned in high-school—a slow, steady accretion of knowledge, building on past successes of scholarship—has proven to be less than fully accurate. The actual process, particularly as defined by Thomas Kuhn, involves a series of fits and starts and trips down dead-end paths before a startling breakthrough—a process more aptly described as revolutionary than evolutionary. Kuhn described this process as a "paradigm shift," most visible when a new idea (often with radically different implications from its predecessors) sweeps across a field, forcing people to completely restructure the very way they think about their work and the problems they attempt to solve. These paradigm shifts, when they occur, stand as the most important steps in the development of new knowledge.

Paradigm shifts occur unexpectedly, and they seldom emerge from a program that was explicitly designed to create them. Copernicus created a new model of the solar system, placing the sun, rather than the earth, in the center, as a simple exercise in understanding the behavior of the planets and stars. Yet the implications of the "Copernican Revolution" extended eventually into almost every phase of human life by forcing a rethinking of the roles of science and religion. Galileo's observations using a simple telescope showed that Venus had phases (like our moon), an impossibility under earlier models of the structure of the universe, and then showed moons orbiting Jupiter, proof that the previous earth-centered models of the universe were fundamentally wrong. In the 1920s and 1930s, Edwin Hubble demonstrated (with a much larger telescope) that our own galaxy did not contain the entire universe, but was in fact but one of an improbably large number of galaxies spread across an unthinkably large space, and he thereby changed our own perceptions of humanity's place and role in the universe. Einstein's Special Theory of Relativity stands as the classic example of a paradigm shift, altering the way that people consider the structure and behavior of all matter and energy. Quantum physics created another such shift, leading to the creation of (among other things) the transistor and subsequently the digital revolution in which we are all now involved.

But paradigm shifts are not limited to the sciences. The development of new religions such as Christianity, Islam, and Buddhism represent equally powerful transformations of thought. The invention of silver haloid film created an entirely new kind of art. The World Wide Web has transformed communication in our modern world, a revolution with implications that we barely understand at this point. The U.S. Constitution and the Bill of Rights represent equally powerful political ideas that continue to evolve and spread throughout the world.

How much has the University of Rochester participated in and contributed to this process? In some ways, of course, we do not yet know, since the full import of an idea rarely emerges immediately. (Galileo's ideas, for example, took centuries before their full consequences were felt.) But we know with certainty that University of Rochester scholars have produced vitally important discoveries and insights that have led to truly paradigm-shifting ideas. These ideas span the entire spectrum of human thought, through the sciences, engineering, social sciences, and the humanities. Some of them transform the way other scholars carry out their work. Some transform the way the educational process itself takes place, and some literally create new fields of inquiry that had not previously existed.

This collection of essays provides some exciting examples of the types of transforming work that have taken place at the University of Rochester through its first 150 years. Obviously, these essays do not include *every* major idea and invention arising from the University, but they do represent a sampling of the cornucopia of great ideas that the University has become. In reading these essays, all students, alumni, faculty and staff (past and present), the leaders of the University, and the donors who have helped to facilitate this work with their generous and vital contributions can take immense pride and pleasure in being part of this wonderful process. Perhaps nothing in the University's past and future can and will better exemplify our motto of "ever better" — *Meliora!*

—*Charles E. Phelps*

Acknowledgments

The editors are grateful to the deans and director of the various University divisions—Lowell Goldsmith, Thomas J. LeBlanc, Charles Plosser, Sheila Ryan, James Undercofler, and Philip Wexler—who participated in the process of selecting the (necessarily limited) topics for inclusion in this collection of essays.

Thanks also to Theodore M. Brown, professor of history and of community and preventive medicine, for his suggestion that sesquicentennial activities include a book memorializing some of the major intellectual contributions of the University's faculty through the years.

This project also was made possible by the gifted, professional work of John Blanpied, the author of many of these essays and whose editorial suggestions greatly improved all of them.

Our thanks, finally, to University of Rochester Press for its good services, counsel, and enthusiasm for publishing this work in commemoration of the University's sesquicentennial.

1

Inspiring America's Composers*

"It will take the life of some man to do it. . . . The director of your school will have to breath fire into a great machine and endow it with his own enthusiasm for a great cause."
—Howard Hanson, letter to Rush Rhees, 26 January 1924

In the flush of his long career Howard Hanson was widely considered one of the most influential of American composers. The wunderkind from Wahoo, Nebraska was the first American to win the prestigious Prix de Rome, in 1921, when he was twenty-four. His opera, *Merry Mount*, was the first truly American opera; the fifty curtain calls it received at its 1934 premiere at the Metropolitan is still a house record. Among a lifetime of prestigious awards was the Pulitzer Prize in 1944 for his Fourth Symphony. He composed over one hundred works for orchestra, chorus, chamber, and solo instruments.

But more than his music-making caused George Eastman and University of Rochester President Rush Rhees to tap him as director of the new Eastman School of Music in 1924. Hanson was only twenty-seven at the time, but had been a full professor at the College of the Pacific in California since he was nineteen, and its dean since twenty-one. He had a reputation as a conductor, a champion of American music, and an administrator with serious ideas about the need for a professional school in a university setting, uniting creators, performers, scholars, and educators under one umbrella—precisely what Eastman and Rhees were

*This essay is adapted, with the author's kind permission, from chapter three of Andrea Kalyn's Ph.D. dissertation, "American Composers' Concerts and Festivals of American Music," University of Rochester, Rochester, N.Y.

looking for.[1] During his tenure as director (he retired in 1964), Hanson
added a doctoral degree in performance and composition, broadened
the curriculum, established the Eastman-Rochester Symphony Orchestra (made up of the first chairs of the Rochester Philharmonic Orchestra and the best of Eastman student performers), and in general guided
the Eastman School of Music into its role as one of the premiere univer-
sity-based music schools in the country. During that time, and until he
died in 1981, he continued composing and conducting.

But Hanson's greatest legacy, apart from shaping the Eastman School
itself, is his championship of American music. As a composer he knew
first hand the frustrations of getting his music performed. Nationally,
the few serious attempts to program new American music had all foun-
dered. Nevertheless, Hanson was intent on building an audience capable of supporting and appreciating American music. George Eastman
obviously agreed, his passion for the "democratization of American
music" reflected in the Eastman Theatre and the orchestra he furnished
with an unusually large number of native-born musicians.

In devising the series of American Composers' Concerts that would
become the core of his work at the Eastman School, Hanson sought to
establish a forum in which American composers could hear their orchestral works. To him, this meant the work of composers born in the
United States or of "established" immigrants who adopted wholeheart-
edly American values. In reality, he favored a kind of European-based
Americanism: European structures, American themes. The modernist
works of new immigrants like Schoenberg, Stravinsky, and Bartók were
never performed as part of the American Composers' Concerts or Festivals of American Music.

The patronage of George Eastman allowed Hanson to carry out his
plans with relative ease. Yet Hanson realized that the success of his venture would ultimately depend on a receptive audience. He began preparing that audience as soon as he arrived in Rochester, playing heavily
on a sense of civic duty nurtured in the city by George Eastman. In his
first official statement as director of the Eastman School, Hanson ex-
uded patriotic enthusiasm:

> No music of a country can be truly great, vital, and sincere unless it
> springs from the personal convictions of the people themselves. That is

the reason that the most important thing in American music today is the development of men to write America's music.[2]

In November 1924, Hanson launched a series of articles in the Rochester *Democrat and Chronicle*. In each of the eight installments he addressed a different aspect of musical composition. Even the title of the series—"Modern Music and its Problems"—reflects his care to establish a bond with his readers and future audience. He was careful to define, in the very first article, his intended audience as the layman—the "average lover of music"—and even more careful to include himself in this category "as one of the champions of the 'lowbrows' by admitting my adherence to the emotional concept of music."[3]

American Composers' Concerts

On 15 January 1925 Hanson announced that the Eastman School of Music would "show its belief in American composition in a practical way by offering young composers of ability opportunities of hearing their own orchestral works." He went on to outline the plan for the first American Composers' Concert:

> American composers are invited to submit manuscript scores of orchestral works not previously performed and not exceeding eighteen minutes in length . . . and three or four works will be selected for each program. . . . These works will be rehearsed and performed under my direction by the Rochester Philharmonic Orchestra in morning concerts which will be free to the public and to which representative music critics from other music centers will be invited. The composers of the works to be performed will be invited to attend the rehearsals and the performance of their compositions as guests of the Eastman School of Music and at the expense of the institution. . . . Of the works performed at these concerts those which appear the most worthy will be recommended for repetition the following season in the regular series of the Rochester Philharmonic concerts and will be recommended to other orchestras for performance.[4]

Musicians across the United States applauded the announcement. Newspapers and musical magazines noted the opportunity afforded young composers, and conductors, aware of the financial obligation, coveted the Eastman School's generosity and predicted new growth in American orchestral music. Frederick Stock, conductor of the Chicago Symphony Orchestra, for example, hailed "the beginning of a new era in American orchestral composition," and Rudolph Ganz, conductor of the St. Louis Symphony Orchestra, praised the idea as "probably the most important step made in the direction of assisting American composers."[5]

Fifty-four composers* submitted manuscripts before the 15 February deadline, among which Hanson and a small committee selected for performance orchestral works by Adolph Weiss, Mark Silver, Bernard Rogers, William Quincy Porter, George McKay, and Aaron Copland. Despite Hanson's assertion that "sometimes you'll give a prize to a work you dislike because it's so well built," all but one of the works selected conformed to his own highly conservative taste. Only Copland's *Cortège Macabre* demonstrated innovative compositional technique; its intensity, polyrhythms, and dissonance made it the most striking work of the group. And even though one critic expressed admiration for Hanson's "disregard of public appeal," this was by no means the case. Hanson valued audience reaction highly and claimed that its judgment seemed "in a great majority of cases to be critically sound, giving [him] new confidence in a natural audience reaction." Copland's *Cortège* may have been the greatest "discovery" of the first concert, for instance, but Hanson did not program the work again until the final concert of the Festival of American Music series, in 1971. Rogers' *Soliloquy*, on the other hand— a beautiful, highly conservative, and easily comprehended work—reappeared on seven concert programs. In fact, Hanson programmed more works by Rogers than by any other composer in the entire series, including himself.

At the expense of the Eastman School, Hanson had invited four prominent New York critics to the first concert, thereby giving it immediate national prominence and attention. The critics responded with guarded enthusiasm, but the general tone was encouraging. Harry Osgoode (the *Musical Courier*) assured Hanson that "this particular concert would take its place as something very distinct in the history of

---

*Or forty-eight: there is some dispute.

American music." Olin Downes (*The New York Times*) later reiterated the significance of the series:

> So far as we know, there is no exact precedent in America, if anywhere, for this admirable procedure. Native orchestral composition has been patronized, encouraged, indulged, rewarded in constructive ways in this country, but we do not know of a plan for stimulating orchestral composition which parallels the educational value of promise of the concerts now organized in Rochester.[6]

Such critical exposure stimulated even more interest, so that within a few years, most major newspapers and music journals across the country regularly reviewed the American Composers' Concerts and, later, the Festivals of American Music.

Hanson staged two concerts the first season, one in the fall, one in the spring, and added a third concert the following year. Though they were drawing over eight hundred people, the Eastman Theatre, which seated thirty-four hundred, virtually swallowed the audience. Hence, in December 1926, Hanson moved the concerts to the smaller Kilbourn Hall. Six years later, rapid growth in attendance demanded that he return the concerts to Eastman Theatre, and by 1935, audiences filled even that hall. One astonished critic exclaimed:

> Those who reiterate that the public does not like American music should visit Rochester sometime during an American Festival and observe that public taking that music to its arms![7]

The American Composers' Concerts were rightly praised for the high, professional quality of the performances. Although the Eastman School and the Rochester Philharmonic Orchestra were fiscally distinct, they shared the same premises, Eastman faculty held most of the principal chairs in the orchestra, and Hanson served as an ex officio member of the board of directors. Hence, he was free to "borrow" both the full RPO and its chamber orchestra, the Rochester Little Symphony, for the concerts.

Hanson was ever attentive to his audience. After the first concert he took one critic's advice and began programming only one hour, rather than two, of new music, and repeated it after the intermission. Audiences reacted favorably, and Hanson maintained the format in subsequent years. In 1927, beginning with the fourth concert, Hanson included ballot slips with each program so that audiences could vote for their favorite works. Allegedly on the basis of these tallies, a jury—inevitably chaired by Hanson—selected works for future performance. Although new, previously unperformed works continued to comprise the core of the concerts, Hanson began almost immediately to repeat favorite works in an effort to develop an American musical canon.

To accommodate the growing number of submissions, Hanson gradually increased the number of American Composers' Concerts per season—a record high of five for the fifth anniversary of the series in 1930, thereafter settling to a standard three or four concerts per season. Meanwhile, as both the American Composers' Concerts and the Eastman School flourished, the demand for performances of student works increased. In the fall of 1936, Hanson inaugurated a separate series of informal noncompetitive student symposia. These semiannual symposia, held before the fall American Composers' Concert and the spring Festival of American Music, were essentially reading sessions by the Rochester Civic Orchestra (and after 1942, the Eastman-Rochester Orchestra), giving student composers an unprecedented opportunity to hear their works performed by a professional orchestra.* The student symposia flourished, and by 1964 Hanson had conducted more than eight hundred student works for orchestra.

Festivals of American Music

The first major change in the programming came in May 1931. In honor of the Eastman School's tenth anniversary, Hanson expanded the twenty-second American Composers' Concert—the last of the season—into an ambitious week-long Festival of American Music. Whereas the American Composers' Concerts featured orchestral music performed by the RPO, the festival showcased diverse genres performed by various

---

*Playing for the student symposia also established the orchestra's reputation as an ensemble exceptional in its ability to sight-read contemporary music.

ensembles. Thus, members of the Eastman faculty directed student chamber, vocal, ballet, and band ensembles in the first two concerts of the festival (including, among others, Hanson's own "Romantic" Symphony and his "Lament for Beowulf" for chorus and orchestra). Hanson directed the RPO in the usual American Composers' Concert on the third evening, and brought the festival to a close the following evening with a gala program of ballet and opera. Thereafter, each concert season ended with a similar festival, although until 1947 Hanson replaced the concluding opera with a fully choreographed American ballet.

The largest festival of the entire series, the seventeenth, was held in celebration of the Eastman School's twenty-fifth anniversary in 1947. It encompassed eight concerts and introduced band concerts into the roster. Thereafter, the Eastman Wind Ensemble, initially conducted by Frederick Fennell and later by Donald Hunsberger, became a consistent part of the festival schedule throughout the 1950s and 1960s. Also in 1947, Hanson programmed his opera, *Merry Mount,* for an NBC broadcast, and afterwards began to incorporate staged chamber operas regularly into the festivals. By 1950, opera performances had completely supplanted the ballets.

The thirty-second Festival of American Music, held in May 1962, emerges as the pinnacle of Hanson's efforts. In honor of the fortieth anniversary of the Eastman School, he organized a series of symposia and presentations to complement the festival's concerts. The program was entitled "Creativity," and each day of the festival was devoted to manifestations of the creative spirit: drama, opera, the humanities, science, the arts, the jazz idiom, and the symphony. The festival opened with the world premiere of *The Great Rage of Philip Hotz*, by European playwright Max Frisch, and throughout the week included symposia featuring literary critics Erich Heller and Northrup Frye, mathematician Samuel Eilenberg, scientist Rene Dubos, art historian Horst W. Janson, and musicologist Edward Lowinsky. Musically, too, the 1962 event was the most elaborate and yet representative of Hanson's festivals, encompassing every aspect of American music included in previous American Composers' Concerts and Festivals. At the same time, it was the most musically diverse, including a program of Latin American chamber music, some "experimental" music, and even a jazz concert, featuring New York's Modern Jazz Quartet in an "Illustrated History of Jazz" and the Eastman Wind Ensemble in three jazz-inspired works by Eastman alumni Jack End and Alec Wilder.

Variations

Alongside this flourishing enterprise of concerts and festivals of live American music, Hanson was busily developing offshoots of considerable importance to the culture of American music in general.

Publication

After the first American Composers' Concert, critics urged the Eastman School to join the effort to change the situation for young composers, who found it virtually impossible to get their orchestral works published. The financial security of the Eastman School in the 1920s again put Hanson in an enviable position, and in 1926-27 the school published its first American orchestral works: Rogers' *Soliloquy* and Porter's *Ukrainian Suite*, both drawn from the first American Composers' Concert. Hanson had cited the publication of "worthy" compositions for performance and study as one of four requirements for developing American composition, yet he recognized that most professional orchestras could not risk programming esoteric, "difficult" works. Hence, all of the forty-three works published by the Eastman School over the course of the series reflected the moderate tastes of the "average lovers of music" who comprised Hanson's audience.

Broadcasts

In 1930, the Columbia Broadcasting Network aired the fifth anniversary concert over its fifty-two stations nationwide. The same year, the Eastman School of Music began broadcasting regularly over the NBC network performances by school ensembles, and the following year made history with an international broadcast featuring works from that season's American Composers' Concerts. By 1936, NBC featured the Eastman School in regular, weekly forty-five-minute broadcasts. During the 1941-42 season, Hanson produced twenty-two programs of American music nationally broadcast over NBC under the title "Milestones in the History of American Music."

Recordings

The Eastman School began recording its concerts and some rehearsals in 1933 for the private use of composers, students, and scholars. Although the earliest recordings are sporadic, audio engineers from Eastman's Audio Archives, supported by grants from the National Endowment for the Humanities, have preserved, on tape, recordings from the original instantaneous aluminum and acetate discs. Later, as the technology improved, all the concerts and performances were routinely recorded. As for commercial production, in 1939 RCA Victor issued its first recordings of American orchestral music with the Eastman-Rochester Orchestra performing twenty-two works drawn from the American Composers' Concerts and Festivals of American Music (including Aaron Copland's *Music for Theatre* and Hanson's first two symphonies). Beginning in the 1950s, Mercury Records released a series of recordings from the festivals under its "Living Presence" label, including over one hundred works each by the Eastman-Rochester Orchestra (Hanson conducting) and by the Eastman Wind Ensemble (Frederick Fennell).

Finale

When Hanson retired from the Eastman School in 1964 and Walter Hendl assumed the directorship, the University of Rochester established the Institute of American Music and appointed Hanson its director. Under the Institute's umbrella Hanson continued the Festivals of American Music until 1971. A gala concert was held on the last day of the final festival in honor of his seventy-fifth birthday. It re-presented the music of the first American Composers' Concert from 1925, including Copland's *Cortège Macabre*, and closed the door on the series.

By that time, Hanson had organized the performances of more than two thousand works by eight hundred American composers, including nearly six hundred premieres (and several hundred more if the student symposia are counted). The concerts did include a few "modern" works by composers such as George Antheil and Carl Ruggles. In 1927,

Wallingford Riegger's atonal *Caprice for Ten Violins* received its premiere at an American Composers' Concert, and in 1931, Henry Cowell first performed his "fiendishly difficult" *Concerto for Piano and Orchestra* in the Eastman Theatre. But these were exceptions. The composers Hanson favored for the series—such as Randall Thompson, Deems Taylor, Leo Sowerby, and Bernard Rogers—sought to write mainstream music of "universal significance": beautiful music in formal structures rooted in the Austro-German tradition and familiar to audiences weaned on Strauss and Sibelius. "Nationalist" works, colored by American folk songs, hymns, Negro spirituals, and evocations of the American spirit, also appeared frequently on Hanson's concert programs. The twenty-one women composers whose works were performed in the concerts and festivals also fit clearly into the nationalist camp. As with his American realist contemporaries in other arts, Hanson's agenda was evident: to stem the tide of European modernism in America, and to build and retain an audience for American music.

In 1976 Hanson donated $100,000 in Kodak stock to the Eastman School with the condition that it go to fund the Institute of American Music. The aim of the Institute—now the Howard Hanson Memorial Institute for American Music—was and is to continue Hanson's legacy by the promotion, performance, and documentation of new American music.

Hanson died in 1981; the eighty-fifth birthday concert planned for him in the Eastman Theatre became instead a memorial concert. In its obituary, *Newsweek* called him "perhaps the most influential educator in American music." *The New York Times*, which had been quite sniffy about the grandiose Eastman Theatre in 1921, conceded that Hanson had made Rochester "a boom town for American music." Hanson had indeed fulfilled Lawrence Gilman's 1924 prophecy in the *New York Tribune*:

No one is ever likely to call this composer a mere dreamer of wan and misty fantasies. He is as affirmative as a northwest wind.[8]

—John Blanpied

2

Discovering Progesterone

"What merits of my own can I claim except that I have been wise enough to stay within the range of my abilities and sufficiently industrious to carry on whatever I have begun?"

—*George Corner*[1]

In 1927 George Corner was a distinguished and experienced physiologist and histologist, chairman of the anatomy department, the first faculty member George Whipple had hired for the new School of Medicine and Dentistry. He was also a classical scholar, a medical historian, founder of the medical school's history club, and guiding spirit of the new medical library. Willard Allen was a second-year medical student in the School's second class, a star biochemist, and the first Fellow in Anatomy in Whipple's innovative fellowship program. When they teamed up it was, in Allen's phrase years later, a "flourishing" association, and it made medical history.

For fifteen years Corner had been researching the mammalian female reproductive system. He had made important discoveries about the reproductive cycle, and about the function of the corpus luteum in the ovary. He had deduced the existence of an unknown substance produced by the corpus luteum, and had surmised that its function was to prepare the lining of the uterus to receive and to nourish the embryo. He was ready now for the payoff—laboratory proof of his hypotheses and chemical isolation of the substance—but he needed a biochemist to assist him. Willard Allen, he said later, was "perfect. He never made a mistake." In 1929, they published their report on the discovery of the female hormone progesterone, establishing it, as the name implies, as the sine qua non of gestation.

When Corner started his work in 1913 very little was actually known about the reproductive cycle. No one knew when ovulation occurs, the function of the recurring corpus luteum, or when an embryo implants itself. Even gynecologists had little physiological knowledge of the cycle. They could do surgery and repair injuries, but could not treat functional disorders of the female reproductive system. Corner's work helped to change all that, and laid the basis for many of the advances in reproductive medicine of the future. It led directly to a number of therapeutic uses of progesterone—regulation of the menstrual cycle, reduction of premenstrual tension, relief of postpartum uterine spasms, and others. But its most notable practical application has been the development of the contraceptive pill.

The Stages of Discovery

The corpus luteum

Corner began studying the corpus luteum as a postgraduate assistant, then an intern, in Franklin Mall's pathology lab at Johns Hopkins University in 1913. The corpus luteum ("yellow body") is a hormone-producing glandular structure that develops in a mammalian ovary at the site of a released ovum. When the follicle, or ovum-sac, ruptures to eject the ovum into the oviduct (fallopian tube), the cells of the follicle's lining enlarge to fill the cavity, forming the corpus luteum (so-called because its cells in some species are filled with yellow pigment), which then secretes the female hormone progesterone.

Pigs

Corner's original mission was to help Franklin Mall determine the ages of the early human embryos in his collection, but it set him on quite a different track. From a local slaughterhouse he harvested 128 pairs of ovaries from pregnant pigs. Corner later dismissed the paper he published in 1914, on the corpus luteum "as it is in swine," but the experience taught him lessons about the gross anatomy and structure of the sow's ovaries which were to be of great value to him in later years.

In 1915 he became an assistant professor in the anatomy department of Herbert Evans at the University of California, Berkeley, moving to a life of research "with the comfortable feeling that I had found the way to a career."[2] He immediately set up shop in a local slaughterhouse, with a little lab-shack right on the killing floor, collecting sows' ovaries in minute stages ranging from the onset of estrus, or heat, until one week after. He found out that ovulation occurs spontaneously one or two days after estrus, no coitus required; he could study follicles in detail and recover ova just after their discharge into the oviduct. His objective was a definite description of the origin of the corpus luteum, but for that he would have to wait; meanwhile, he had observed the first part of the mammalian reproductive cycle in the kind of detail never before witnessed in any large domestic animal.

Back at Johns Hopkins as an assistant professor in 1919-20, Corner dramatically extended his research, with the objective of making a full study of the whole reproductive cycle of the pig. Along the way, he made a discovery that has been important for obstetricians and their patients. The prenatal loss of an embryo—miscarriage—was not, as Mall had thought, due to a defect of the uterus or malfunction of the corpus luteum, but rather was caused by some inherent deficiency in either the sperm or the ova.

Corner arranged with a local piggery for a thousand ovaries to be delivered to him via the slaughterhouse. "It took me months, of course, to have thin sections made of the follicles and corpora lutea and uteri and to study them. But finally putting together what I had learned in Berkeley and Baltimore, I had the whole story of the sow's cycle before me." In 1921 he published his report in a monograph that, he later thought, "marks my appearance as an authoritative scientific investigator."

Progestational proliferation

In that paper he introduced the term "progestational proliferation" to describe one of the primary functions of the corpus luteum—that is, to make the uterus welcoming and protective of the embryo-to-be. The beauty of his work, Corner thought then and since, was the completeness with which he was able to describe "the schedule of estrus, ovulation, the fate of the unfertilized ovum, the rise and fall of the corpus luteum, and the progestational change in the uterus."[3] In addition, the scheme of the sow's cycle was so simple and well-defined that it provided a kind of template for those researching the far more confusing cycles of other species.

Monkeys

The next step was to address the puzzling phenomenon of menstruation, which occurs only in humans and some of the higher primates—apes, baboons, old world monkeys. To understand its physiology and place in the reproductive cycle, Corner needed to do a more sustained and controlled study than the occasional human autopsy made possible. Thus began his research on rhesus monkeys. He obtained eleven young females from the Philadelphia Zoo in 1921, and set up the first long-term primate lab in the U.S.*

For nearly two years Corner studied these monkeys in great detail to establish their ovulation and menstrual cycles. He then killed and autopsied them in a regular sequence. The result was the first complete picture of the entire reproductive cycle of the higher primate, including the relation of menstruation to ovulation. "I was able to show that in this species ovulation occurs at the middle of the interval between two menstrual periods, and that the corpus luteum persists, if the egg is not fertilized, about fourteen days, as in the sow." Menstruation occurs at the same point in the cycle as estrus does in pigs, though they are different phenomena. At this point Corner was still unable to relate the onset of menstruation to the state of the corpus luteum, which was his objective, but he did manage to refute a prevailing dictum of the German embryologist, Robert Meyer: "Without ovulation, no menstruation." Corner showed that in young females, anovulary menstruation is indeed possible.

---

*Years later Corner reflected on his thirty-years of work on reproduction in the rhesus monkey. "I studied by observation and experiment approximately five hundred monkeys, each of which was observed for months and most of them for several years. A single experiment with the ovarian hormones often called for two or three months of daily injections and examinations. As a final stage in many cases the ovaries had to be cut into serial sections so that every corpus luteum and large follicle could be examined in detail. My technicians thus sectioned, stained, and mounted on glass slides for me to study at least a hundred pairs of ovaries numbering altogether 200,000 or more individual sections" (*Anatomist at Large*, 47).

Rochester

The climactic stage of Corner's long road of research was to occur in Rochester where, in 1926, "I began to plan an all-out attempt to find the hormone of the corpus luteum."[4] His first task was to establish the physiological basis for the biochemical work. He did so in a paper, the first of the series "Physiology of the Corpus Luteum," in which he described tests on rabbits which confirmed the unique and essential role of the corpus luteum in the survival of embryos for longer than four days.

Enter Willard Allen, the young biochemist whom Corner invited to accept the School's first "year-out" fellowship in anatomy. Beginning late in 1927 they set out to isolate and then to purify a potent hormonal extract. They undertook exhaustive tests using sows' ovaries to find a solvent that would extract the hormone from the corpus luteum tissue. Allen finally managed it with alcohol and the use of an extraction column from the lab of physiologist Walter Bloor.

To test the extract, they devised what came to be known as the Corner and Allen Test. First,

> mate your [rabbit] doe to a fertile male; eighteen hours later remove both ovaries (easier than digging out the corpora lutea separately), inject your extract daily for five days, kill the doe, prepare a cross section of the uterus, put it under your microscope. If the progestational change has occurred, you have a potent extract.[5]

They used young rabbits first to see if the progestational changes occurred in the uterine lining. The ultimate test came with the use of mature rabbits, "to see whether our crude extract would substitute for the mother's corpus luteum so fully as to preserve the embryos." It did, and "there was much excitement in Corner's laboratory on that sunny afternoon."[6] Finally, they "clinched the matter" by carrying several rabbits through to term and the delivery of healthy offspring.

"Physiology of the Corpus Luteum" [7]

In 1929, Corner and Allen announced their discovery in two historic papers. These were parts two and three of the series "Physiology of the

Corpus Luteum," that Corner had begun the year before, establishing the unique and essential role of the corpus luteum in the survival of an embryo.

In the first of the new papers—"II. Production of a Special Uterine Reaction (Progestational Proliferation) by Extracts of the Corpus Luteum"—the coauthors review the findings of paper number one:

> The results recounted [there] . . . led us to attempt the preparation of extracts of the corpus luteum capable of producing progestational proliferation of the endometrium [uterine lining]. At the same time these experiments have placed in our hands a standardized method for testing the extracts.

They went on to describe the Corner and Allen Test, then their experimental methods, results, observations, and the preparation of the extracts, all in exacting detail. Finally they described the way in which they ruled out the possibilities of other accidental sources of the demonstrated potency, showing that they "do not produce the uterine reaction characteristic of early pregnancy and the corpus luteum phase of the cycle." They considered the recent claims by Edgar Allen and Edward Doisy for oestrin (estrogen) as the potent hormone, and concluded that "progesterone proliferation is caused by a specific hormone, differing from oestrin, which is elaborated by the corpus luteum."

Finally, in distinctly understated summary:

> It appears, therefore, that the extracts of corpus luteum contain a special hormone which has for one of its functions the preparation of the uterus for reception of the embryos by inducing progestational proliferation of the endometrium.

In their second joint paper ("III. Normal Growth and Implantation of Embryos After Very Early Ablation of the Ovaries, Under the Influence of Extracts of the Corpus Luteum") it remained "only to determine whether the proliferation thus induced will serve to protect the embryos and bring about implantation."

The authors carefully describe their new rabbit studies, the removal of the ovaries and corpora lutea at different stages following ovulation, the injections of the extract. They then concluded:

These experiments demonstrate that in the presence of progestational proliferation induced by corpus luteum extracts . . . the embryos may survive and grow normally and normal implantation may occur, whereas in the absence of progestational proliferation the embryos never survive beyond the fourth day. The evidence is now complete that in the rabbit the corpus luteum is an organ of internal secretion which has for one of its functions the production of a special stage of the uterine mucosa (progestational proliferation) and that in turn the function of the prolif-erated endometrium is to nourish or protect the free blastocysts [newly fertilized eggs] and to make possible their implantation.

Progesterone

There followed a lengthy process on Allen's part to purify the extract. Finally, in May of 1933, he arrived in Corner's office holding a test tube with some tiny crystals in a solution: "This is it. It's a steroid," said Allen. "What's a steroid?" said Corner.

By 1934 the hormone had gone international. Four competing teams—two American, two German—had achieved almost simulta-neous results in purifying the extract. A competition of naming ensued, which was finally resolved at a London conference called by the League of Nations' Health Division to standardize the values of newly discov-ered sex hormones. Allen went, carrying Corner's suggestion of a com-promise name: neither his preferred "progestin" nor competitor Karl Slotta's "luteo-sterone," but "progesterone." Coming four years before World War II, the conference of ambitious, talented, contentious, but finally amicable scientists stands as a model exercise of harmony.

By 1937 Corner and Allen announced that they had achieved with a crystalline progesterone extract all five of the established effects of the corpus luteum upon the uterus: progestational proliferation, suppres-sion of ovulation, inhibition of the uterine muscle contraction, inhibi-tion of uterine motility in vivo, and suppression of menstruation. The last four of these effects are related to the corpus luteum's function of protecting the embryo once it has been implanted. The inhibition of the muscle contraction, for instance, helps prevent premature labor; the inhibition of ovulation insures that the embryo will not have to compete with a new one for space and nourishment.

Aftermath

Corner continued his work on the primate menstrual cycle with monkeys. In 1927 he had been one of the first to state explicitly that menstruation is brought about not by the positive action of the corpus luteum, but by its decadence. The comparative roles of estrogen and progesterone were, through the 1920s and 1930s, hotly debated on both sides of the Atlantic. By 1939 Corner was able to assemble the various data to form a working hypothesis of the interaction. In brief, it stated that "the uterine lining is kept in normal condition by both the ovarian hormones, estrogen . . . and progesterone. During that phase of the cycle in which progesterone is produced by the corpus luteum, this hormone acquires a kind of dominance over estrogen; thus, when the corpus luteum breaks down, the uterus, now protected by neither of the two ovarian hormones, lapses into menstruation."[8] Published when Corner was fifty-five, this article marked the culmination of his thirty years of work on the female reproductive cycle.

In 1940 Allen began a distinguished career as professor of obstetrics and gynecology at Washington University. Corner returned to Baltimore to direct the Department of Embryology at the Carnegie Institute of Washington (at the Johns Hopkins Medical School), and retired in 1956 to write a history of the Rockefeller Institute for Medical Research. In addition he authored an autobiography, two biographies, a history of the University of Pennsylvania Hospital (writing at Ben Franklin's old desk), medical texts, essays in medical history, translations, and books of advice to girls and boys on attaining woman- and manhood. He became the executive officer of the American Philosophical Society, retiring from that job at the age of eighty-nine. He died in 1981, at ninety-two.

—John Blanpied

3

Of Diet, Drugs, and the Cure for Anemia

"Teaching and research represent the ultimate in pleasure and satisfaction in my career. Perhaps research may give a greater sense of accomplishment, but teaching carries greater personal happiness . . . I would be remembered as a teacher."

—*George Whipple*[1]

At the University of Rochester there are still those who remember George Whipple as colleague, mentor, and teacher. As co-founder of the Medical Center and for over thirty years its head and guiding spirit, he seems embedded in its very DNA. He came to Rochester in 1921 at the behest of President Rush Rhees and George Eastman, with the support and encouragement of the Rockefeller Foundation, to supervise the design, construction, and staffing of a new School of Medicine and Dentistry and the Strong Memorial Hospital. As dean of the School, Whipple conducted his "three-legged stool" of a career with all the granite aplomb of his New Hampshire upbringing. He was an exceptional administrator. Teaching brought him great personal satisfaction. But it was the third leg of the stool, his research, that made his scientific reputation and brought him international acclaim. Through the building of a great medical center, through two world wars and the Great Depression, for fifty years, rigorous and indefatigable, from student to professor emeritus, he conducted his meticulous research on plasma and muscle proteins, bile and iron metabolism, and the life cycle of the red blood cell.

The heart of that research—for which he received the Nobel Prize in Physiology in 1934—is his classic work on blood formation and the role of the liver in the regeneration of red blood cells lost to anemia.

The story of the prize is often confused because Whipple shared it with two Boston physicians, George Minot and Richard Murphy, who contributed quite different kinds of work, though it all led to one outstanding conclusion. According to the Nobel presentation,

> Of the three prize-winners, it was Whipple who first occupied himself with the investigations for which the prize has now been awarded. He began in 1920 to study the influence of food on blood-regeneration, the re-building-up of the blood, in cases of anemia consequent upon loss of blood. . . . These investigations and results of Whipple's gave Minot and Murphy the idea, that an experiment could be made to see whether favourable results might not also be obtained in the case of pernicious anemia.[2]

Anemia in general is a disease marked by a reduction in red blood cells, or in the oxygen-carrying protein, hemoglobin, produced by those cells. But there are many different kinds of anemia. Whipple did his groundbreaking research for the most part on severe secondary anemia, whose cause was known to be the loss of blood. He was chiefly concerned with the influence of the liver and of dietary factors in the body's remarkable ability to manufacture its vital blood products. His work in the 1920s on this form of anemia, and his discovery of the potently beneficial effects of an iron-rich liver diet upon the regeneration of hemoglobin and red blood cells, has contributed importantly to our general understanding of liver and blood functions. But its most direct and outstanding result (thanks here to Minot and Murphy) has been the clear conquest of a deadly disease, *pernicious anemia*, and the saving of many thousands of lives worldwide.

Pernicious anemia, the particular study of Minot and Murphy, is distinct from simple and secondary anemia. It is a "deficiency" disease, meaning that an intrinsic factor vital to the normal division of red blood cells is missing from the body.* But it was Whipple's work that led Minot and Murphy directly to their raw-liver experiments on afflicted patients, with such dramatically successful results that in a very short time what had been a mysterious, incurable, debilitating, and usually

---

*See the discussion later in the essay.

fatal (hence, "pernicious") disease, whose only available therapy was regular blood transfusions, large doses of arsenic, or removal of the spleen, had become routinely treatable.

The Groundwork

Whipple's major work was carried out in Rochester. He had laid the groundwork at Johns Hopkins University as resident pathologist in the great William Welch's lab, where his study of bile pigments led to an interest in hemoglobin production and to the exceedingly complex functions of the liver. Doing research on chloroform-poisoned dogs, for instance, he found that the liver cells had seemingly miraculous powers of regeneration.

At the University of California Medical School (1914-1920), Whipple continued this work on bile metabolism, bile formation, and the relation of liver to blood components. But in 1918 he shifted his emphasis to the direct investigation of the role of the liver in anemia. He and his assistant Charles Hooper, whom Whipple had brought with him from Johns Hopkins, induced artificial anemia in dogs by large, carefully controlled hemorrhages in order to study the rate of blood regeneration in response to a variety of dietary factors. They found that feeding the dogs radically different diets caused radically different blood results: very slow hemoglobin regeneration from carbohydrate-rich diets, very rapid regeneration from a diet of lean meat—beef heart or, best of all, liver.*

A key player in all Whipple's work at that time was his indispensable assistant, Frieda Robscheit-Robbins. Her first job was to manage the great many large dogs used in the research, but she became his true collaborator on the seminal experiments there and throughout the twenties in Rochester. (According to Katherine Whipple, her husband felt

---

*Whipple later recounted how, in 1918, Hooper had injected three pernicious anemia patients with a liver extract. Their condition immediately improved, but the work was ridiculed by clinical doctors and Hooper did not pursue it. Whipple lamented that the discovery of the treatment for pernicious anemia could have occurred ten years earlier than it actually did ("Autobiographical Sketch," 266).

that Robscheit-Robbins should have shared in the Nobel Prize. He did in fact share the prize money with her and with his mother.) By 1920, Robscheit-Robbins's name was appearing with Whipple's on a new series of papers about blood regeneration after simple anemia. These papers made quite clear that liver was a potent blood-forming substance, but also pointed out the need for long-term studies under rigorously controlled conditions, with better methods for monitoring and measuring the volume and contents of the blood. The need for more accurate measuring methods set off a parallel program of research in which Robscheit-Robbins played a prominent role, particularly in the laborious quest for a suitable aniline dye.

Rochester Results

When Whipple was wooed by the irresistible Rush Rhees to come to Rochester to head up the University's new medical center, Robscheit-Robbins stayed behind for a year to carry on the experiments. Meanwhile, even while overseeing the design and construction of the new medical center, Whipple arranged to continue research. He had a two-story research lab built near the construction site, sufficient to house various temporary offices, working labs, and animals. In 1922 Robscheit-Robbins brought her forty Dalmatian-English bull dogs on a train from California, installed them in the new lab, and continued the dog-studies with scarcely a hitch.

And now came the payoff for all their years of careful work. In a series of six papers appearing between 1925 and 1926, collectively titled "Blood Regeneration in Severe Anemia," Whipple and Robscheit-Robbins described their crucial and painstaking experiments. They had developed much-improved methods for obtaining accurate quantitative measurements of blood constituents, and for maintaining the dogs in uniform states of anemia over the course of many months without threatening their safety. (These techniques included "plasmapheresis," in which the dogs were repeatedly bled, then injected with washed red blood cells suspended in a protein-free medium. The protein-free medium maintained the anemia while preventing the animals from suffering shock.) In this state, the dogs were fed a seemingly infinite variety of diets and drugs under rigorously controlled conditions. This program produced steeply varied rates of red cell and hemoglobin regeneration, all now measurable to a high degree of accuracy.

"Blood Regeneration in Severe Anemia"[3]

The first paper in the series explains the "standard basal ration bread" diet of the dogs, and the experimental methods used throughout the experiments. The clarity and thoroughness of the paper itself demonstrates both Whipple's experimental style and the cast of his mind:

> Our earlier experiments in this field have been concerned with simple anemia in dogs produced by two or three large hemorrhages at the beginning of the anemia period. . . . These recent experiments . . . deal with a *constantly maintained severe secondary anemia*. An attempt is made to reduce the hemoglobin level to about 40 or 50 per cent and to maintain this anemia by frequent bleedings of calculated amounts. The bleeding samples are measured for total hemoglobin and we are able to state from week to week how much hemoglobin the body can produce over and above the maintenance factor. . . .

He confronts the question: Why did they use dogs in the experiments? And answers: rabbits and guinea pigs won't eat meat and can't metabolize it; rats have inaccessible veins and too little circulating blood to bleed frequently or measure accurately. "Dogs are omnivorous and on suitable diets can be kept in health over many months in spite of extreme experimental anemia." The dogs were maintained in strictly controlled conditions of comfort, and cleanliness, conditions obviously essential to the success of the experiments.*

The paper then goes on to describe in lucid detail and very plain English the methods of feeding, bleeding, and drawing the samples, of

---

*In 1946 two of the Dalmatians—Josie and Trixie—received an award from Friends of Medical Research "on behalf of their ancestors who twenty years before had contributed to the fundamental research which led to the discovery of the liver treatment for pernicious anemia" (George W. Corner, *George Hoyt Whipple and His Friends: The Life-Story of a Nobel Prize Pathologist* [Philadelphia, 1963], 210).

measuring the plasma, hemoglobin, and red cell hematocrit values, of generating the analytical indices, and preparing the basal bread diet.

The second paper in the series is the keystone. Of this paper—"Favorable Influence of Liver, Heart and Skeletal Muscle in Diet"—Whipple's biographer, George Corner, remarks: "This report with its unequivocal emphasis on liver feeding is the most important single paper as regards George H. Whipple's world reputation as a scientist, in the whole of his immense lifetime list of more than 300 publications."[4] Whipple begins unambiguously:

> Liver feeding in these severe anemias remains the most potent factor for the sustained production of hemoglobin and red cells as indicated in various tables. This favorable and remarkable reaction is invariable in our dog experiments no matter how long continued the anemia level, no matter how unfavorable the preceding diet periods may be and regardless of the substances given with the liver feeding.

Later, having detailed the experiments and the results, he asks: "How may we explain this remarkable productive reaction of hemoglobin and red cells due to liver feeding? In our opinion this is evidence that the *body stores in the liver* parent substances which are used in the construction of hemoglobin and red cells." It is a theme he was to develop in his research over the next decade and more.

In the third paper, "Iron Reaction Favorable," Whipple revisits an earlier conclusion on the vexed and long history of the role of iron in anemia and sounds a characteristic note of caution:

> The history of anemia treatment with drugs is indeed a tale to make the judicious grieve. . . . Our earlier experiments with short anemia periods gave no evidence that iron treatment was of the slightest value. . . . *But* subsequent experiments with long standing *severe anemia* give just as positive evidence that iron is of considerable value. These experiments may well emphasize the need of caution in deductions drawn from slightly dissimilar experimental conditions.

And he ends with a conclusion that has in fact stood the test of time:

> We should not lose sight of the fact that under all our varied experimental conditions in secondary anemia due to hemorrhage, *food factors* (es-

pecially various meat products) are most favorable for rapid regeneration of hemoglobin and red cells. It is probable that in human beings *food factors* will be found more efficient in the control of simple anemia than iron or other drugs. Even in complex anemias (human pernicious anemia, anemia with nephritis and cancer cachexia) food factors deserve serious consideration in the clinical management of the blood condition.

In subsequent papers—twenty-one in all until 1930—Whipple and Robscheit-Robbins go on to describe the exhaustive tests of other foods and drugs. Green vegetables—surprisingly, even spinach—proved to be mostly inert when it came to blood regeneration; most fruits, except dried apricots, disappointed; dairy products, and notably whole milk, were useless. Animal organs were unquestionably the best food for the blood: liver (except fish liver) reigned supreme, followed by pig and beef kidney, then brain and pancreas, though spleen tissue and even bone marrow (despite its being the very tissue for forming red blood cells) were of little effect. Of drugs and minerals, iron in some cases was found helpful, and useless in others; but liver in all cases was the unchallenged king.

Some of the papers dealt with attempts to isolate and fractionate the active ingredient in liver, though success was limited at this point. The question remained as to whether the organ contained a single chemical agent apart from iron which, consumed as food, stimulated hemoglobin production in the anemic dog. Finally Whipple was forced to conclude that, at least for the secondary anemias he studied, no single factor was responsible for the remarkable beneficial properties of a liver diet, but rather a complex array of protein derivatives, minerals, and amino acids.[*]

Pernicious Anemia

These reports caught the attention of Minot and Murphy in Boston, and gave them the idea of treating human victims of "primary" pernicious

[*]The later discovery of Vitamin B_{12} as the single-agent therapy for pernicious anemia did not surprise Whipple. He had already confirmed that for secondary anemia due to hemorrhage the liver furnished a broad and complex array of properties. For a while, Eli Lilly, the drug company, marketed a separate drug for secondary anemia. When that proved too confusing, a combination drug, Lextron, was produced.

anemia with the liver diet that had been so successful with the anemic dogs. (The dogs may have been happier with the treatment. In the case of the human patients, who had already lost their appetites, it meant the ingestion of half a pound of raw liver a day.) Minot and Murphy's methods involved careful microscopic monitoring of the blood samples, since pernicious anemia is famous for its deceptive temporary remissions. Within just two weeks, their first group of forty-five patients all showed significant improvement.

The success of the liver diet confirmed that pernicious anemia was a *deficiency* disease, caused by the body's lack of a so-called "intrinsic factor," a term coined by the physician William B. Castle in the 1920s. In healthy bodies, as we now know, this intrinsic factor (a glycoprotein secreted by the stomach) combines gastrointestinally with a vital "extrinsic factor" supplied by the diet (Vitamin B_{12}). The complex protects the vitamin from degradation in the stomach, until it can be released in the ileum of the small intestine. From there, it is absorbed and goes on to do its crucial work in the division of red blood cells, the production of the protein hemoglobin, and the diffusion of oxygen to the tissues of the body. Where the intrinsic factor is lacking, as in pernicious anemia, inadequate amounts of the iron-bearing vitamin are absorbed into the body, which is then starved of oxygen.

In the late 1920s, Minot and Murphy's experiments, following Whipple's, proved that the extrinsic factor, still unidentified, resided in liver. Their success set off a new search for a high-potency extract of the vital ingredient in whole liver. In 1928, George Minot and the chemist Edwin Cohn isolated "fraction G" and made it available to Whipple for tests. Taken orally, the extract remained the sole treatment for pernicious anemia until 1948 when Karl Folkers in America and Alexander Todd in England simultaneously succeeded in isolating and identifying the therapeutic anti-anemia factor as the iron-bearing Vitamin B_{12}.

During the 1930s and 1940s Whipple entered into a mutually profitable collaboration with Eli Lilly and Company whereby his university lab conducted tests for the company. He took out no patent royalties on the products he and his assistants had developed (which was the then hotly debated practice of some other medical institutions), but the company paid fees that both augmented the pathology department's research budget and built up a handsome reserve fund. By the time of

Whipple's retirement in 1955, the fund had grown to $700,000, sufficient to endow the chair of pathology, create a number of scholarships for entering students, and to provide for visiting lectureships.

It was not until 1971 that the great organic chemist Robert Burns Woodward succeeded in synthesizing Vitamin B_{12}. Thereafter, it proved so much cheaper to produce than the liver extract and so much easier to administer, by tablet or injection, that Whipple's essential contribution to the chain of events ironically began to drop out of the literature on the subject, a victim of its own success.

Nevertheless, for more than twenty years after winning the Nobel Prize, Whipple continued his wide-ranging research on bile salts and pigments, on the participation of iron in red cell formation, on protein and metabolism, and half a dozen other related subjects. His sustained studies of iron metabolism and distribution led, in 1934, to his formulation of the theory of Dynamic Equilibrium of Proteins, resulting in a book of that title in 1956, and, in George Corner's estimation, laying the basis for the whole modern theory of protein metabolism.

At eighty-one (seventeen years before he died in 1976), Whipple hoped to be "remembered as a teacher," and generations of his students will probably fulfill that hope. Institutionally, the University of Rochester may be most indebted to him as shaper and administrator of its medical school. The rest of us, aware of it or not, are the beneficiaries of a remarkable career in medical research.

—John Blanpied

4

Breakthrough Chemistry:
The First Synthesis of Morphine

"There are more or less efficient ways to do it, but first you have to have the idea."

—*Marshall Gates*

In 1952, two years after coming to the University of Rochester at age thirty-five, Marshall Gates announced a landmark achievement in organic chemistry: the first successful laboratory synthesis of the complex alkaloid compound, morphine. Recognized as a masterpiece by organic chemists throughout the world, Gates's elegant work sparked an explosive quest among chemists and pharmacologists for a nonaddictive derivative of the medically important drug, brought Gates himself international repute, and helped make the University one of the country's leading centers of research in organic chemistry.

In the early 1950s, the "art and science" of organic synthesis was in its infancy. Robert Burns Woodward of Harvard (who would win the Nobel Prize in 1965) had synthesized the complex alkaloid, quinine, during World War II; William S. Johnson of the University of Wisconsin, also working with very large molecules, would shortly synthesize the female steroid hormone estrone; and Gates's work on morphine was (according to Columbia University Professor Gilbert Stork) "the first multistep total synthesis of any complex natural product." Gates was way ahead of the field, one of a small group of pioneers who ushered in the golden age of organic synthesis which has characterized the second half of the twentieth century.

Organic Chemistry

Organic chemistry is that branch of chemistry concerned with the structure, behavior, and properties of carbon-based compounds, which are by far the largest class of compounds in the world. Carbon itself is unique in the variety and extent of structures that can result from the three-dimensional connections of its atoms. When carbohydrates, produced by photosynthesis, combine with variable amounts of hydrogen, oxygen, nitrogen, sulfur, phosphorus, and other elements, the structural possibilities of the resulting compounds are limitless.

The Big Idea, the foundation of the science of organic chemistry, is the late-nineteenth-century structural theory which explains how carbon and other elements combine in unique, particular, three-dimensional arrangements to form the staggering diversity of organic compounds. This theory is "one of the most successful qualitative generalizations in all of science. . . . Its principles have not only correlated an extraordinary mass of observations but have required few fundamental additions during the last hundred years. The classic structural theory of organic chemistry is of the same importance in the history of science as the development of the two laws of thermodynamics around 1850, the quantum theory and the theory of general relativity after 1900, the explanation of the molecular basis of genetics after 1950, and the plate tectonics ideas in geology."[1]

The corollary to the structural theory sounds disarmingly simple: that the property and behavior of a substance are determined by its structure. Simply put, to understand the natural world, it is necessary to understand the structure of its parts, the architecture of its molecules. Organic chemistry seeks to do that. And at the heart of organic chemistry is the "fine art and exact science" of organic synthesis, whose leading motive is to assemble complex molecules from simple pieces. If the synthetic target is a naturally occurring substance, the successful synthesis is the final proof of a molecular structure that has been proposed through analysis. Once synthesized, the compound can be recreated in the lab, or it can be modified.

Organic Synthesis

To the extent that any legal drug you take is 1) available, 2) effective, 3) safe, that is probably because it's an unnatural derivative of a natural substance—in other words, the work of the modern organic chemist. Fabrics, perfumes, explosives, and pesticides, too, are almost all synthetic these days.

There are two preliminary stages before the actual synthesis. The first is the isolation and purification of the natural product from its host organism. Morphine, for instance, was isolated from the opium poppy by the German apothecary F.W.A. Sertürner in 1806. Next comes the process of "degrading" the substance: breaking it down into ever-more-basic, recognizable, constituent parts, until the functional groupings—the three-dimensional arrangements of the atoms—are understood.* (Modern instrumental techniques can reveal the structural details for amounts weighing as little as one-millionth of a gram.) At this point, owing to the methods, the observations, and the logic used, the chemist may be confident of the chemical structure of the molecule. And yet it remains a hypothesis, a proposal, a suspended possibility, until the final stage of the synthesis itself, which is no less a project than creating the molecule in the lab—rebuilding it from scratch. Once built, if the newly constructed compound proves to be identical to the natural one, then the synthesis is successful, and the molecular structure has been confirmed.

The obvious challenge for synthetic chemists is the mind-boggling structural complexity of most organic substances, especially of the larger molecules. Synthesizing morphine, for instance, is like facing a huge three-dimensional jigsaw puzzle whose atoms must be pieced together in the proper spatial relationships. Or, to twist the metaphor, "just as a given pile of lumber and bricks can be assembled in many ways to build houses of several different designs, so too can a fixed number of atoms be connected together in [a multitude of] ways to give different molecules. Only one structural arrangement out of the many possibilities will be identical with a naturally occurring molecule."[2]

*This "degrading" is the classical method of ascertaining a structure. Today structures (at least of crystalline substances) are determined by x-ray crystallography in most cases—a process involving hours instead of months or years.

The reasons for accepting the challenge will be various and probably multiple. First, as noted earlier, the synthesis is the ultimate proof of a hypothesized structure. Then there is the sheer fact of the intellectual challenge itself—the chemist's version of the mountain-climber's lure: "because it's there." Beyond that, the demands of the structure often require improvements in our knowledge of chemical reactions and chemical properties—that is, the chemists may have to develop new chemical reactions and understanding, and come up with new ideas of how to make complicated molecules. As craftsmen, they will be able to test and demonstrate their recondite skills. And as theorists, they may test new theory by designing new molecular structures to see if, and how, they work. On the practical side, the synthesized version may be cheaper or more easily accessible than in its natural form. More importantly, the synthetic methodology enables the chemists to ring changes in the structure, to add, subtract, rearrange, or substitute an atom or functional group here or there, to make a new structure that behaves differently.

Ringing the changes, whether for directly practical applications or for theoretical research, is the heart of the organic chemists' craft. The synthesizers work on both natural and unnatural materials. In the early days of organic synthesis it was enough to recreate the products of the natural world to learn their whys and hows and wherefores. But in the last few decades, with the spectacular advances in instruments and in related fields like biochemistry and quantum mechanics (revealing the mechanics of chemical reactions in living systems and the mysteries of chemical bonding, for example), the lion's share of the work is done on unnatural materials. Dyes, for instance, once were all derived from naturally occurring substances. Now those used in clinical work or in neurochemical or biochemical research as well as in rugs and foods are far more likely to be synthetic.

The organic synthesizers *make molecules*—either natural or unnatural—but probably they will tell you that making substances not from nature but from their own imaginations is at the heart of what they do, the raison d'être of their craft. After that, if the new thing finds a practical, benevolent application, well that's fine too. But first and foremost, synthesis is a creative act.

Everyone, Marshall Gates included, acknowledges Robert Burns Woodward as the greatest organic chemist of the century, brilliantly worthy of his 1965 Nobel Prize for his "outstanding achievements in the art of organic synthesis." So numerous and original were his achievements—steroids, alkaloids, antibiotics, resperine, chlorophyll, Vitamin B_{12}, and on and on—that his legacy is the knowledge among organic chemists that, given enough time, money, and manpower, they can make anything. That legacy is shared by Marshall Gates. Before the work of those pioneers in the late 1940s and early 1950s no one really knew, not only *how* to synthesize complex organic compounds, but to what extent it was even possible. Now, so swiftly did the intellectual revolution come, and so completely did it sweep through and transform the field, it may be difficult to recognize the early uncertainties.

Morphine

Since its isolation from the opium poppy in 1806, morphine has been unsurpassed in its power to reduce the level of physical distress without the loss of consciousness, and has therefore long been considered among the most important of naturally occurring compounds.* Though partly supplanted for medical use by some of its modern derivatives, it is still used today to treat pain caused by cancer, and in cases where other analgesics have failed. It also has a calming effect that protects the system against exhaustion in traumatic shock, internal hemorrhage, and congestive heart failure. Alas, aside from such undesirable effects as depression of the respiratory, circulatory, and gastrointestinal systems, it is of course highly addictive. Among its more famous derivatives are heroin and a number of analgesics, including codeine.

As the first alkaloid to be isolated and crystallized (1806), morphine is the prototype of the class, and because of its well-known uses and effects, it has long been a highly provocative target. Its structure was correctly proposed in 1925 by the English chemist Sir Robert Robinson.

---

*Morphine is an alkaloid, a large class of naturally occurring nitrogen-containing bases. Other well-known alkaloids include strychnine, quinine, ephedrine, and nicotine. Because of their diverse physiological effects, alkaloids have excited interest since ancient civilizations, but have had to await the growth of organic chemistry for their intricate structure to be unraveled.

But though its constitution had been identified, and though considerable research was carried out and hundreds of papers written on the subject over the next quarter century, the ultimate synthesis remained an alluring challenge.

Gates

Marshall Gates received his Ph.D. at Harvard in 1941, and spent two years at Bryn Mawr before going to Washington with the National Defense Research Council (NDRC). There, over lunch, he took to pondering the intriguing problem of morphine synthesis. All synthesis starts with an idea, he said later. With no laboratory available, and certainly no computers, unable to do any chemistry, working in his head and on paper, he conceived a method to selectively construct the principal skeleton of the molecule with all its constituent atoms in their correct three-dimensional places. With this structure, he reasoned, he ought to be able to manipulate the functional groups on the skeleton's periphery, so as to alter the basic structure until he achieved the arrangement assigned to morphine by Robinson in 1925.

After the war, back in the lab, Gates carried out his conception of the synthesis, first at Bryn Mawr and then, with ultimate success, at Rochester. The procedure took twenty-six steps, all but one of which Gates carried out himself, with his own hands, without the laboratory instruments that were shortly to revolutionize the entire field of organic synthesis.* Gates's solo method—somewhat unusual even at the time, and virtually unknown today — was partly a function of the small labs both at Bryn Mawr and Rochester, and partly the working-habit picked up as a graduate student at Harvard, with Louis Fieser, who, though he supervised a large group of researchers, nevertheless continued to do hands-on research throughout his professional life.

Every account of Gates's work mentions its intellectual "elegance" and his virtuoso craftsmanship. Gates himself is more modest. That he

---

*The one step, the "resolution," was done with his postdoctoral student, Gilg Tschudi, who also prepared the degradation product from the morphine alkaloid thebaine (important as an intermediary in the synthesizing process).

was and is a glassblower, for instance, he looks upon as an efficiency for a chemist; but his colleague Jack Kampmeier, an organic chemist in the Rochester department, sees it as characteristic of the man who designed, in his head, one of the most creative, interesting, and aesthetically pleasing experiments of modern chemistry, and carried it to completion with "brilliant hands."

Consequence and Context

Gates's achievement brought him worldwide attention, numerous distinguished lectureships, and, among other honors, election to the National Academy of Sciences. It also brought prominence to the chemistry department at the University of Rochester, and inspired an era of important research, locally and internationally, into the development of nonaddictive derivatives and new synthetic analogues of morphine and other opium alkaloids. Gates himself, through the 1950s and 1960s, synthesized, tested, and patented hundreds of related compounds. He witnessed the introduction of instrumentation (nuclear magnetic resonance, infrared spectroscopy, and others) and the methods and insights of molecular biology, biochemistry, and quantum mechanics. Designs impossible to conceive without better three-dimensional information, or procedures too laborious to undertake without rapid and accurate tools for determining molecular structure, became possible, and the "exact science" of organic synthesis was radically changed and improved. Yet it remains true that synthesis starts with an idea, and that Gates's idea about morphine was the fountainhead for a widening flood of discovery in synthetic organic chemistry.

The University of Rochester chemistry department in the 1940s and early 1950s was small but on its way toward a national reputation. W. Albert Noyes, Jr., son of a distinguished scientist, and a prominent physical chemist and photochemist in his own right, had come from Brown University in the late 1930s with a mandate to create a first-rank department. The young Dean Stanley Tarbell, a Harvard Ph.D. fresh from a postdoctoral year at the University of Illinois, arrived at the same time. Tarbell's mission was to build up the then-minuscule organic chemistry program, to which end he was joined during the next decade by Virgil

Boekelheide, and, in 1949, by Marshall Gates. Meanwhile Noyes, besides being a famously energetic administrator and researcher, was also editor of *The Journal of the American Chemical Society* (JACS), the premier journal in the field. He had known Gates at the NDRC during the war, and asked him now to come to Rochester to handle the organic and biochemical aspects of JACS. Gates agreed on condition that he be given a lectureship and the use of a laboratory to continue his work on the synthesis of morphine. The arrangement suited everyone. As Tarbell said later, "No department could be second-rate with Marshall Gates on its faculty."

In Tarbell's view, the 1950s were the golden days of high intellectual excitement among collegial faculty and excellent students and postdocs doing top-quality research. New instruments seemed to appear every day, the labs were bubbling, government funding was plentiful, and Ph.D. chemists were in hot demand by American industries. Marshall Gates modestly puts the University of Rochester chemistry department among the nation's top twenty in those days; others rank it a good deal higher. Certainly, thanks to Gates, and also to Tarbell and Boekelheide, the department was a national leader in organic synthesis. In 1953, William Saunders joined the organic team to build the newer field of physical organic chemistry, the modern chemistry of reaction mechanisms that provides the theoretical basis for organic synthesis.* Between 1952 and 1961, the department awarded 124 Ph.D.'s, which was twenty percent of the University-wide total, employed 79 postdoctoral students, and published over three-hundred research papers. And, unique among the University departments then and now, three of its faculty—Noyes, Tarbell, and Gates—were members of the National Academy of Sciences (NAS).† The work in the 1950s established a "school of organic synthesis" at Rochester that was propelled forward by the work of Andrew Kende, Richard Schlesinger, and Robert Boechman. Today the

---

*Tarbell's years witnessed "a great increase of knowledge regarding the steps through which organic reactions proceed, the mechanism of reactions. Reaction mechanism work was a major component of physical organic chemistry, which . . . became one of the most popular areas of research, partially eclipsing the classical fields of synthetic and structural organic chemistry" (Dean Stanley Tarbell, *Autobiography*, 151).

†Boekelheide was also elected to the NAS after he went to the University of Oregon in 1960.

Department of Chemistry maintains its strong tradition of synthetic organic chemistry, though it has also grown in other directions: physical organic chemistry (the science of reaction mechanisms), photochemistry, chemical physics, biophysical chemistry, and, starting in the 1970s, a whole program of inorganic chemistry.

In tandem with his advancing academic career, Gates maintained his involvement with JACS as assistant editor till 1962, as editor-in-chief until 1969. He was by common consent a brilliant editor, exercising the sensitivity, taste, intellect, and scientific wisdom that marked his masterful work in the laboratory.

In 1968 Gates became Charles Frederick Houghton Professor of Chemistry. In 1981, at the age of sixty-five, he was forced into retirement by inexorable university laws that were to be liberalized only one year later. He continued to work in the lab for several more years. In 1995, to mark his eightieth birthday, the Upjohn Company established a $200,000 Marshal Gates Faculty Scholar Award at the University. A colleague, Jack Kampmeier, spoke on the occasion of the "world-class intellectual and technical achievement" of Gates's famous synthesis, calling it "a triumph of scientific imagination."[3]

—John Blanpied

5

The Rochester Conferences on
High Energy Physics*

"[The conferences were memorable for the] genuine collegiality
that prevailed, the informal discipline accepted by the participants
. . . and the subtle ways in which the entire roster . . . served as a
global planning group for . . . research activities in particle physics."
— *Robert E. Marshak*[1]

In 1945, a unique generation of physicists dispersed from their wartime
assignments and resumed their peacetime lives. They did so as acknowl-
edged war heroes, whose intellectual prowess had quite literally won
the war. A grateful citizenry granted them an extraordinary privilege—
the opportunity to use the powerful tools they had helped create during
the war effort to explore nature, limited only by their talent and imagination. In the words of the late Nobel laureate physicist, Luis Alvarez,
"Right after the war we had a blank check from the military because we
had been so successful . . . we never had to worry about money."[2]

This euphoric state of affairs could not last. Inevitably, competition
from other national priorities, including competing scientific priorities,
coupled with the increasing cost of the quest undertaken, would con-
spire to bring this historic moment to a close, leaving the field that had
been created—high energy physics—but one of many worthy competitors for public support. But before this transition to "business as usual"
took place, the foundations for what has come to be called the Standard

*The sections of this essay describing the Shelter Island and Rochester conference
series are drawn from R. E. Marshak, "The Rochester Conferences: The Rise of International Cooperation in High Energy Physics," *Bulletin of the Atomic Scientists* 26
(1970), and J. C. Polkinghorne, *Rochester Roundabout: The Story of High Energy Physics* (New York, 1989).

Model had been laid. This extraordinary theory describes an underworld beneath the realm of atoms and nuclei populated by exotic entities—quarks, gluons, and W & Z bosons—no one could have predicted. As a result, our knowledge of our physical environment has been forever enriched. Remarkably, the name of the University of Rochester is permanently associated with this unique moment in the history of physics thanks to the vision of Robert Marshak, whose "Rochester Conferences" came to serve as the official chronicle of this period.

Prologue

Marshak's appointment to the faculty in 1939 was to prove a pivotal decision on the part of physics department chairman, Lee DuBridge. DuBridge had been hired by President Rush Rhees in 1934 with the mission of transforming the department from one primarily focused on teaching undergraduates into a leading center for research and graduate study in physics. His first step toward accomplishing this goal was to hire Sidney Barnes, an experimental physicist with exceptional technical skills. Together, DuBridge and Barnes would quickly catapult Rochester to the front ranks of nuclear physics by building one of the nation's first cyclotrons.

Thanks to the generous patronage of George Eastman, the University of Rochester was in a very favorable financial position in the 1930s. This enabled DuBridge to make other superb faculty hires as well, including Fred Seitz in 1935 and Victor Weisskopf in 1937.* However, his plans for developing the department would be interrupted by the advent of World War II.

Rochester physicists played active roles in the war effort. Foremost among these were the contributions of DuBridge himself, who served as director of the MIT Radiation Laboratory, arguably the most important of the scientific and engineering enterprises assembled in the United

---

*DuBridge, Marshak, Seitz, and Weisskopf would all be elected to the National Academy of Sciences. DuBridge would become president of Caltech, Marshak president of City College of New York, Seitz president of Rockefeller University, and Weisskopf director general of CERN, the great postwar European high energy physics laboratory.

States during the war years. DuBridge was a remarkably effective administrator, and he guided the "Rad Lab" in developing the important advances in radar and long range navigation that played such key roles in determining the war's outcome.*

The physics department faculty also played important roles in the Manhattan Project, where Weisskopf earned the sobriquet "the Los Alamos Oracle" in an environment of which it has been written: "It is probably safe to say that never before in the history of the human race have so many brilliant minds been gathered together in one place."[3] While at Los Alamos, Marshak developed a theory of radiative diffusion of heat waves that resulted in their becoming known as "Marshak waves," and devised new methods for studying the crucial problem of neutron diffusion in matter.

Birth of a Golden Era

The physics faculty who returned to Rochester after the war did so to a very different research environment than the one they had left. Vannevar Bush's seminal document "Science—The Endless Frontier," commissioned by President Roosevelt in 1944 and delivered to President Truman in 1945, was to provide the basis for national scientific policy during the Cold War era. It would lead to the establishment of the Atomic Energy Commission in 1946, charged with supporting basic research in nuclear and high energy physics, and to the creation of the National Science Foundation in 1950.

The resulting influx of generous federal support for basic research led to a period of explosive growth in the U.S. scientific enterprise. As a result, there were numerous postwar options for the returning generation of physicists to consider. Weisskopf relocated to MIT, and DuBridge himself left Rochester in 1946 to become president of Caltech, where he went on to oversee the development of an institution second to none in science and technology. But fortunately, DuBridge had remained in Rochester long enough after the war to secure funding from the Office

---

*DuBridge would later observe that "[t]he atom bomb only ended the war. Radar won it" (D. J. Kevles, *The Physicists* [New York, 1978], 308).

of Naval Research for the construction of the University's second particle accelerator, the 130-inch synchrocyclotron, which Barnes brought into operation at the end of 1949. This accelerator, second in capability only to Berkeley's 184-inch machine, was to put Rochester "on the map" in postwar high energy physics.

To the department's lasting good fortune, Marshak chose to remain at Rochester, where for the next twenty-five years he devoted his great talents and prodigious energy to building a major physics department, while carrying out research in theoretical physics that would bring him international renown, including election to the National Academy of Sciences in 1958 and presidency of the American Physical Society in 1983. In 1970, Marshak left Rochester to become eighth president of the City College of New York.

The Shelter Island Conferences

The physicists returned from the great wartime collaborations with more than just their government's goodwill. The war had greatly advanced the state of many technologies, such as those involved in the use of microwaves, which would quickly find application to the new accelerators and particle detectors that would enable rapid progress in high energy experimentation. But equally important, these physicists returned with a first-hand appreciation of the power of collaborative effort. The wartime teams that had been assembled had solved problems of immense difficulty and complexity, and they had done so in a remarkably short period of time. This represented a marked departure from the way in which physics research had traditionally been carried out. The paradigm of the individual genius working alone in the lab—Ernest Rutherford and a lab assistant discovering the atomic nucleus in 1911— had been superceded by one involving teams of specialists united in the construction and exploitation of large common facilities.

Theoretical physicists also recognized the value of intellectual collaboration, and after the war they too sought opportunities through which they could benefit from direct interaction with each other. Out of this came a sequence of three annual conferences organized by Oppenheimer, held successively in Shelter Island, N.Y. (1947), Pocono Manor, Pa. (1948), and Oldstone-on-Hudson, N.Y. (1949). These meetings were small, each involving about twenty-five participants, largely theoretical in nature, including at most two or three experimentalists to serve as "expert consultants," and almost exclusively American. Their programs were largely, although not exclusively, focused on resolving certain outstanding questions in the theory of quantum electrodynamics. After the third meeting Oppenheimer concluded that the original goals for the conferences had been achieved and he terminated the series.

Marshak had participated in the Shelter Island conferences, and had experienced first-hand the catalytic effect of intense group discussion with colleagues at the highest level of scientific sophistication and imagination. In particular, a discussion at the first Shelter Island Conference had stimulated him to make one of his most insightful contributions to high energy physics—exemplifying the creative potential of the interactive environment that Marshak would succeed in duplicating in his own series of conferences.

The Two-Meson Hypothesis

The impressive success of quantum electrodynamics—the theory of the interaction of photons and electrons—had encouraged physicists to seek an analogous theory of the strong nuclear force, which binds protons and neutrons within the atomic nucleus. In 1934, Japanese physicist Hideki Yukawa proposed that instead of being massless like the photon, the carriers of the nuclear force might have mass, which would provide a natural explanation for the rapid decrease of the nuclear force with distance. Yukawa's theory, for which he would later be awarded the Nobel Prize, predicted that the mass of these hypothesized carriers of the strong nuclear force would need to be about two-hundred times that of the electron to be consistent with what was known about the range of nuclear forces. The discovery in cosmic rays, only a few months later, of a new particle having precisely the mass predicted by Yukawa appeared to provide striking confirmation of his theory, except for one major difficulty—the new particles reacted only very weakly with nuclear matter and for this reason could not be the carriers of the strong nuclear force.

At Shelter Island, Marshak proposed a way out of this puzzling situation by suggesting that the strongly interacting particles predicted by Yukawa were indeed produced in cosmic rays, but only high in the atmosphere, where they promptly decayed into the weakly-interacting particles observed at the surface of the earth. This was a bold hypothesis, unsupported at the time by any direct experimental evidence, but it was proved correct when British scientists, using photographic emulsions carried aloft in balloons, found direct evidence for the decay process Marshak had postulated.

The Rochester Conferences

In 1950, Marshak became chair of the Rochester physics department, and immediately moved to fill the void created by the termination of the Shelter Island conferences. On 16 December 1950 he organized a one-day meeting in Rochester that was to be the first of an influential and continuing series of conferences devoted to the field of high energy physics. Fifty people assembled for this occasion, funding for which was provided by local Rochester industry (including Haloid Corporation, soon to be renamed Xerox).

Marshak's new conferences differed in important ways from the Shelter Island series. Most significantly, he believed that "a new series of conferences should be inaugurated in which the experimentalists should be given 'equality' with the theorists."[4] This was a most timely decision since the rapidly evolving field of high energy physics was to be driven by experimental discoveries, often in directions that the most brilliant theoretical physicists had not anticipated. The other hallmark of the Rochester Conferences was their international character. Marshak correctly sensed early on that high energy physics was not only a collaborative enterprise, but that the teams involved were going to become increasingly international. Soon, being invited to the next Rochester Conference would become one of the most sought-after indicators of worldwide recognition in the field.

The first seven conferences were held in Rochester, the first five in mid-winter (not usually thought of as Rochester's tourist season). Nevertheless, the conferences immediately caught on, with a list of attendees that constituted a veritable "Who's Who" of the physics community of the era. The third conference attracted federal support and acquired a more formal title—"Third Annual Conference on High Energy Nuclear Physics"—while the fourth meeting had the sad distinction of being the last Rochester Conference attended by Enrico Fermi, who died later that year at the early age of fifty-three. The fifth conference saw the first major influx of foreign participants, with the attendance of representatives from fifteen countries. In another move toward internationalization, this meeting was cosponsored by the International Union of Pure and Applied Physics (IUPAP), which would henceforth play an increasing role in the organization of the Rochester Conferences.

Achieving this level of foreign participation was no simple matter, and how this was accomplished provides an interesting glimpse into the political times. This was the McCarthy era, and to make matters worse, the government had recently revoked Oppenheimer's security clearance. As a result, Oppenheimer's presence at the conference introduced political overtones, and the McCarren-Walter Immigration Act, designed to exclude communists from the United States, was invoked to exclude the participation of some European physicists. Showing his usual energy and enterprise, Marshak contacted his congressman (and future senator) Kenneth Keating, and together they went to Washington and convinced the State Department to sponsor the necessary waivers from the U.S. Attorney General.

Perhaps it was the increased participation of foreign physicists, who wanted to see more of Rochester than just the inside of Bausch & Lomb Hall, but for whatever reason, the sixth and seventh conferences were held in April, a considerably more hospitable time of year in upstate New York. These two meetings were noteworthy for more than just the improved weather—they included one of the famous incidents that made the Rochester Conferences not just forums at which new results were presented, but more like international working-group meetings from which totally unexpected ideas could emerge.

The Discovery of Parity Violation in Weak Interactions

One of the great mysteries of the early 1950s, and a topic of discussion at all the early Rochester Conferences, centered on the properties of certain new particles that had been discovered in 1947 in cosmic rays. Two types of particles, more massive than those with which Marshak had been concerned at Shelter Island, had been observed: q-particles, which decayed into two charged particles, and t-particles, which decayed into three. What was mysterious about these particles was that, except for their different decays, they appeared to be otherwise indistinguishable, with identical masses, lifetimes, and spins. Nevertheless, it could be shown that, because of their different decays, q's and t's could not be the same particle if the weak nuclear interaction responsible for their decay preserved right-left symmetry, a property with the technical name "parity conservation." Parity conservation was widely viewed as one of the fundamental laws of nature. However, at the sixth Rochester Conference, in a memorable session chaired by Robert Oppenheimer, Richard Feynman posed a question, which he attributed to experimentalist Martin Block, challenging this assumption: "Could it be that the q and t are different parity states of the same particle which has no definite parity, i.e., that parity is not conserved? That is, does nature have a way of defining right- or left-handedness uniquely?"[5] In the ensuing discussion, C. N. Yang made a comment that before long would take on added significance, observing that "he and [T. D.] Lee looked into this matter without arriving at any definite conclusions."[6]

Feynman and Murray Gell-Mann, who had also participated in the discussion, returned to Caltech "very excited over the question of parity non-conservation,"[7] but they had only a few weeks to ponder the matter before Lee and Yang submitted to *Physical Review* their definitive paper on the subject. Based on a review of the experimental situation, they concluded that while "existing experiments do indicate parity conservation in strong and electromagnetic interactions . . . for the weak interactions . . . parity conservation is so far only an extrapolated hypothesis unsupported by experimental evidence."[8] They then went on to identify some possible experimental tests that could resolve the matter. Shortly thereafter, experiments would confirm that parity was indeed violated in weak decays. Thus, by the time of the seventh Rochester Conference, parity violation had gone from a provocative question raised at the previous meeting to an established fact, and, later that same year, the basis for Lee and Yang's receipt of the Nobel Prize.

The International Conferences on High Energy Physics

In response to the increasingly international character of the Rochester Conferences, it was decided by the IUPAP High Energy Commission that, beginning in 1958, the conferences would be renamed the "International Conferences on High Energy Physics," but be numbered consecutively from the first Rochester Conference. Henceforth, the conferences would rotate among different global venues, with successive conferences being held in the United States, Western Europe, and the Soviet Union. Since this was the height of the Cold War, to include the Soviet Union on this list was a very conscious choice, and reflected Marshak's steadfast commitment to the goal of creating a truly international high energy physics community.

The eighth and ninth conferences were held in Geneva, Switzerland, and Kiev, USSR, respectively. On every dimension, these two meetings were poles apart in organization and execution. The Geneva meeting introduced a new level of professionalism into the series, including handsomely bound proceedings, although at some cost to the "cozy family atmosphere"[9] of the old-style Rochester-based meetings, while the Kiev meeting was obviously somewhat of an ordeal for the participants. But only by bringing the meetings to Russia could physicists such as future Nobel laureate Lev Landau be provided an opportunity to participate.

The last Rochester Conference actually held in Rochester was the tenth, in August 1960, when about three-hundred physicists from thirty countries assembled on campus. The value of having held the previous conference in Russia was reinforced when at the last moment the entire Russian delegation to this conference withdrew without explanation. The attendees included Nobel laureates and laureates-to-be in abundance—a notable photograph from the conference shows eight Nobelists standing in front of the Delta Kappa Epsilon fraternity house. Marshak's creation had come of age and was about to leave home for good. However, notwithstanding their peripatetic future travels among the major cities of the world—the Thirtieth International Conference on High Energy Physics took place in the summer of 2000 in Osaka, Japan— the conferences have never forgotten their birthplace. In informal discourse, wherever in the world they are held, they remain the "Rochester Conferences."

—Paul Slattery

6

The Biopsychosocial Model

"The physician's basic professional knowledge and skills must span the social, psychological and biological, for his decisions and actions on the patient's behalf involve all three."

—*George Engel*[1]

What It Is, How It Happened

George Engel proposed his biopsychosocial model of medicine in an article in *Science* in 1977 entitled "The Need for a New Medical Model." The argument was deceptively simple: that illness and disease are always caused by a combination of biological, psychological, and social factors, whose relationships affect both the process and the outcomes of care.

> The dominant model of disease today is biomedical, with molecular biology its basic scientific discipline. . . . [But] to provide a basis for understanding the determinants of disease and arriving at rational treatments and patterns of health care, a medical model must also take into account the patient, the social context in which he lives, and the complementary system devised by society to deal with the disruptive effects of illness, that is, the physician role and the health-care system. This requires a biopsychosocial model.[2]

The biopsychosocial model was not new, and had in fact been in development and use in Rochester for thirty years at the time. It was the outgrowth of a set of educational initiatives that Engel and John Romano brought to the School of Medicine in 1946, which over the years transformed both the practice and spirit of medical education at the School, and played out in both clinical practices in the hospital and the community, and in medical research. The corps of medically trained liaison fellows in psychiatry and internal medicine (the Med-Psych Liaison Unit), the Program in Biopsychosocial Studies, and the fellows in family medicine and family therapy are all modern offshoots of Engel's model. In research, it laid the groundwork for the Division of Behavioral and Psychosocial Medicine and the interdepartmental Center for Psychoneuroimmunology Research, headed by Robert Ader, George L. Engel Professor of Psychosocial Medicine. It remains today a major part of the medical school curriculum, providing its philosophical underpinning even in an age towering with immense investments in diagnostic and technological and managed-care health-delivery systems.

Engel himself was trained as an internist in the no-nonsense behavioral biomedical science tradition at Johns Hopkins and Mt. Sinai in the 1930s, dismissive of anything smelling of the then-gathering winds of psychosomatic research and theory. But at Peter Bent Brigham in Boston, and later at the University of Cincinnati with the young psychiatrist John Romano in the early 1940s, he made a number of observations that led him to abandon his resistance to the psychological factors in medicine. Working among colleagues in neurophysiology and psychiatry, he undertook "collaborative research in which he now explored in imaginative and open-ended ways the psychological as well as the medical dimensions of his clinical cases."[3] He and Romano also conceived dramatic new strategies of medical education by which psychiatry would be integrated into every level of the medical curriculum, psychosomatic workshops would be taught, and experienced psychiatrists would be employed as teachers and co-workers rather than just as occasional visitors. To implement their ideas Romano and Engel sought and gained grant support from the Commonwealth Fund and the Rockefeller Foundation for a new psychosomatic training program.

Nothing stiffens departmental barriers like the threat of curricular changes, and in the end, Cincinnati proved resistant to the proposed changes. In 1946, John Romano came to the School of Medicine and Dentistry of the University of Rochester for the challenge and opportunity of chairing the newly instituted Department of Psychiatry. Engel followed with a joint appointment in medicine and psychiatry and a special mandate from Romano to plant the Cincinnati ideas in

Rochester's more welcoming soil. They brought the Commonwealth and Rockefeller funding with them, and set about their revolution with energy and dispatch.

The Right Place, The Right Time, The Right Stuff

"Rochester couldn't have chosen a more appropriate time [than 1946] to launch a department of psychiatry in a modern medical school in the United States," Romano wrote later.* It was a period of federal largesse, the expansion of health insurance programs to include the mentally ill, and the establishment of psychiatric units in general teaching hospitals. And significantly, "in the spirit of the postwar period, there was a pervasive interest in providing the physicians of the future with more systematic knowledge of man in human and social terms as well as in the traditional biological context of medicine."[4]

Engel was immediately impressed with the medical school's spirit of interdisciplinary cooperation, the "permeable" boundaries between disciplines. In William McCann and the dean, George Whipple, he sensed "commitment and dedication to the school as a whole" rather than the parochialism of separate departments as famously found in larger, older, more exalted medical schools. He was especially impressed with the School's commitment to the students themselves—a most unusual priority, in his experience, among medical schools. (He soon discovered the meaning of George Corner's iconic phrase, "the other end of the log"—a reference to the collegial intimacy between faculty and students.) A medical school small enough that everyone, students, nurses, junior and senior faculty, and staff, all ate together in a common room, was one conducive to something "different and innovative."

The Medical Curriculum

Romano and Engel saw as their primary mission not the education of psychiatrists, but the development of broadly prepared humanistic phy-

---

*A field of little interest at the medical school through 1945, psychiatry had been housed in the Department of Medicine, chaired since the beginning by William McCann.

sicians. To that end, they mounted an energetic assault on the curriculum. Romano instituted a new required course for all first-year students in "Fundamental Concepts of Human Behavior," while Engel began with one for second-year students in "Psychopathology," learning on the go in a relatively new field for him.* At the same time he developed an elective "Psychosomatic Clinic" for third- and fourth-year students and house officers, and integrated medical rounds for the third-year students. Engel's Rockefeller Foundation fellow began offering a Clinical Conference in the Medical Outpatient Department for third- and fourth-year students focusing on the more common emotional problems of patients which are met in the daily practice of medicine.

Throughout the 1950s and 1960s, Romano and Engel enhanced the program with a series of robust and ambitious modifications, moving always toward broader and more rational integration of the biopsychosocial perspective into the entire undergraduate curriculum.

One of Engel's most important innovations was his second-year course in medical interviewing, taught three or four afternoons a week during the second semester. Very little attention had been paid at Rochester, or anywhere else, to the business of taking a patient's history, a boring and unpopular duty which students were expected to pick up "out of their hip pockets." The ability to glean significant psychosocial information pertinent to the patient's distress was considered more or less a function of the student's, and then doctor's, general humanity, compassion, and intuition. Yet Engel saw the patient interview as a serious procedure, in fact *the* crucial feature in the patient-physician relationship, and central to the whole biopsychosocial concept of medical care. Under guidance of faculty mentors, small groups of second-year students interviewed real patients. During these formal sessions, and informally on the wards of Strong Memorial Hospital, students were mentored by post-residency fellows working in the Med-Psych Liaison Unit, top-quality graduates of residency programs in internal medicine, pediatrics, and obstetric/ gynecology. Over time, Engel managed to rewrite the school-wide format for the patient history.

---

*Within a couple of years he'd developed enough confidence to begin distributing the course notes to his students, and these notes became the basis for his groundbreaking 1962 textbook, *Psychological Development in Health and Disease*.

In 1975, John Romano described thirty years of distinctions in the Department of Psychiatry, concluding that "first and foremost, the department has been recognized for its devoted and inspired teaching of its medical students throughout the floors and clinics of the Hospital."[5] A few years later, Engel reflected on the distinctive nature of the Rochester education. Typically in medical schools, he observed, students learn *content* in the first two years, not principles and process. At Rochester, in the psychosocial aspects of their training, students learn how to be *physicians*, and notably how to elicit real information—crucial biopsychosocial information—from their patients. Throughout their training they develop greater knowledge and skills in "people science"— all learned, Engel emphasized, not haphazardly, incidentally, or sentimentally, but in a systematic orderly way. Indeed, the success of the integrated curriculum was the widespread acknowledgment that psychosocial training is as susceptible to scientific methods as any other.[6]

Medical-Psychiatric Liaison Unit

The liaison program conceived in Cincinnati was Romano-Engel's most far-reaching innovation in Rochester at the graduate level, and put into place with the enthusiastic cooperation of medicine chairman McCann. The liaison fellows were recruited after a minimum of two years' residency in medicine, pediatrics, or obstetrics-gynecology, then very intensively supervised on wards, where they were able to add "psychological parameters" to their well-established medical training. Later they were assigned to medical students as tutors and joined in the third-year medical liaison rounds and the fourth-year outpatient teaching, as well as in intensive research seminars and conferences. As the program developed, members of the liaison unit also participated in teaching exercises in several of the community hospitals, and several intern and resident members of the community hospitals were assigned to the liaison service activities at Strong Memorial Hospital.

After about five years the Liaison Unit became a formal part of the Department of Medicine. (Later on, a comparable Behavioral Pediatrics Unit was established in the Department of Pediatrics.) For the most part faculty in the other departments welcomed the Med-Psych fellows into their programs—having recognized, as Engel says, that it was possible to stay in medicine and still merge with psychiatry. One obvious reason for the credibility of the program, among colleagues as well as students, was that, unlike similar programs elsewhere, the liaison officers—Engel himself, William Green, Franz Reichsman, and Arthur Schmale—were all trained internists.

The hallmark of the now much-imitated program was, and is, the thorough integration of real-patient clinical research with teaching. "Through the extensive teaching activities of these units and their interdepartmental research activities, the influence of Romano, Engel, and their colleagues began to be powerfully felt throughout the institution. They became leaders in articulating the educational agenda, and they had a profound effect in shaping institutional culture."[7]

Monica

One of the most famous, and exemplary, research projects of the 1950s illustrates much about the climate and methods of the Romano-Engel program. Engel cites his fortuitous introduction to the child Monica as an example of the "permeability of barriers" at the school. A nurse at the common lunch table mentioned this "interesting child" in pediatrics with a gastric fistula and puzzling emotional problems—did Engel and Franz Reichsman care to take a look? They did so, and thus launched a landmark study of the relationship between behavior and gastric secretion. They discovered, for instance, that in the presence of strangers, Monica lapsed into an extreme "depressive-withdrawal" condition, whereas among familiars, and especially with her favorite, Reichsman, she would be lively, responsive, and happy. In each instance, her gastric activity was characteristically different and fully integrated with her total behavior.[8]

Serendipity therefore took Engel and Reichsman into pediatrics, a wholly new experience, and subsequently into gastroenterology and surgery. Over the years Monica, endearing herself to the researchers and students alike, became a major vehicle for medical education, an opportunity for teaching child development and for learning how to observe a child. Thirty years later, in 1983, students were still observing

Monica, whose name had become famously associated with a classic research project and a whole new field of psychoanalytic research "through which visceral processes throw light on mental events which could not be understood otherwise."[9]

Monica provides an example as well of Engel's teaching philosophy—that the patient is our teacher; that regardless of specific issues the well-studied patient provides opportunity for learning general principles. "Any patient is of interest," he said, encouraging teachers to work in areas where they weren't necessarily expert—internists in surgical service, for instance—and so to discover what is common to all of medicine, regardless of discipline.

Legacy

"It's like God, it's everywhere," a student said of the biopsychosocial model at Rochester. Today it has been thoroughly integrated into the undergraduate curriculum—in the longstanding year-long first- and second-year course in Psychosocial Medicine, the one-hundred hours every student puts into medical interview training over the first two years, in the general clinical and specialty "clerkship" rounds of the third and fourth years, and in such new courses as the Introduction to Human Health and Illness, an effort to link the biomedical and psychosocial domains through a series of "whole case conferences." It runs through the new Double Helix curriculum, a "problem-based-learning" model of undergraduate medical education in which first-to-fourth years are all interwoven with real clinical study, and in which all courses are interdisciplinary. It informs the Division of Medical Humanities, the Community Outreach and International Programs. The biopsychosocial model is certainly rooted in the affiliations of the University with the several area teaching hospitals, most especially the Family Medicine Programs at Genesee and Highland Hospitals. And like God, the Engel language is everywhere too—in the school's own literature, for instance, proclaiming Rochester "home of the biopsychosocial model," famous for its tradition of the "biopsychosocial integration of the curriculum and the learning experience."

Research

Once the educational innovations had been well-established in the School, Engel could turn more of his attention to his parallel interests in psychosomatic research, both his own and in the department generally. One extremely fruitful theme, initiated by Engel, Robert Ader, William Green, Arthur Schmale, and others in the 1950s, related to biopsychosocial effects of separation, loss, bereavement, and depression. Schmale, for instance, conducted a two-year study showing a relationship between "psychic reactions to unresolved loss" and "changes in biological activities" leading to a variety of physical disorders.

The most revolutionary playing-out of the biopsychosocial approach in basic research, however, came with a momentously serendipitous discovery by Robert Ader in 1974. In the course of a fairly standard Pavlovian conditioned-response experiment with rats, Ader encountered some unexpected and, at first, utterly baffling results. Having conditioned the rats to associate saccharine with the nausea-inducing drug, Cytoxan, Ader then offered them the unadulterated sugar-water to see how long their aversion would last without the actual addition of the drug. Apart from the initial nausea, the rats were not supposed to suffer physical symptoms. Mysteriously, however, they began to sicken and die, and, investigating, Ader learned that one of the Cytoxan's "side effects" was as an immune suppressant. The conclusion seemed as clear as it was astounding: the rats had been conditioned to associate the saccharine not just with nausea, but with shutdown of their immune systems. To put it dramatically—but it *was* dramatic—their minds had been controlling their bodies.

Engel and Romano had recruited Ader in 1957 as a young experimental psychologist whose interest in psychosomatic animal studies coincided with their own sense of the future of the field. Only in a climate of "permeable" disciplinary boundaries could Ader's work have been done—certainly not within immunology itself, where the autonomy of the immune system was considered axiomatic. Ader was and is a conservative scientist, and he conducted repeated tests before he was ready to announce his discovery to the world. Resistance nevertheless was widespread, and it took Ader and his team years of continued demonstration to make their case irrefutable: that the immune and the central nervous system operated as an functionally integrated defense system.

In 1981 the neurobiologist David Felten, soon to join Ader in Rochester, made actual sighting of a hard-wired connection between the systems. Among the results of their joint efforts (along with immunologist Nicholas Cohen and other team members) is, today, the Center for Psychoneuroimmunology Research of which Ader is director. In its first two decades the young discipline has, in Engel's words, "uncovered unsuspected relationships between the two most complex systems that have evolved to assure the survival of the individual and the species."[10] The hybrid name itself (psychoneuroimmunology) was coined by Ader in 1980 and used to title his 1981 book. About 600 pages long, the book was an attempt to collect what little was being done in the field at the time. An indication of the explosive growth since then is that in 1991 the second edition (coedited with Felten and Cohen) ran to 1200 pages, and the third edition, due out in 2001, will consist of three volumes.

Psychoneuroimmunology—"the study of the interactions among behavior, neural and endocrine function, and immune system processes"[11]—is perhaps the quintessential disciplinary hybrid envisioned by the biopsychosocial model. And it is fitting that upon Engel's retirement in 1979, the George L. Engel Professorship of Psychosocial Medicine should be established, thus officially acknowledging the discipline, and that its first occupant should be Robert Ader.

The implications of psychoneuroimmunology for understanding and treating disorders from autism to AIDS to cardiovascular diseases are profound. The practitioners and teachers of the biopsychosocial model of medicine may speak of "psychosomatic disease," but the word is in fact a vestige of a discredited dogma of body-mind dualism. They argue that there is really no such thing as a psychosomatic disease, since that implies that there are also diseases in which psychological and social factors play no part—a view they roundly reject.

If the human body is truly an integrated system, says Ader—and it is—then we will be forced to consider new definitions of disease itself. Compartmentalized kinds of disease—mental, autoimmune, endocrine, gastrointestinal—show us our own limits, not the body's. The germ theory is not enough, the biomedical model is not enough. All diseases are in fact determined by particular mixtures of multiple social, psychological, and biological factors. Reductionist medicine tends to avoid, ignore, or suppress these variables. But in the biopsychosocial model, they are the very heart.

—John Blanpied

7

Writers and Editors: Preserving the Canon

"[I]t is the premise of this edition that Emerson's first thoughts, the hardheaded things written for himself only, the personalia, sometimes even the false starts and unfinished sallies of thought, as well as the things which did go into the essays—all these are real facts to be valued. . . ."

—*William H. Gilman*[1]

In early February of 1851 John Nichols Wilder, one of the University of Rochester's founders, escorted a distinguished literary visitor to the new institution just as it was settling into its first home on West Main Street. Impressed by the energy of the enterprise but also amused by its ambitions, the visitor, Ralph Waldo Emerson, later observed in his journal,

[T]he University of R . . . was extemporising here like a picnic. They had bought a hotel, once a railroad terminus depot, for $8,500, turned the dining room a chapel by putting up a pulpit on one side, made the barroom into a Pythologian Society's Hall, & the chambers into Recitation rooms, Libraries, & professors' apartments, all for $700. a year. They had brought an Omnibus load of professors down from Madison bag & baggage, Hebrew, Greek, Chaldee, Latin, Belles Lettres, Mathematics, & all Sciences, called in a painter, sent him up a ladder to paint the title "University of Rochester" on the wall, and now they had runners on the road to catch students. One lad came in yesterday; another, this morning; "thought they should like it first rate," & now they thought themselves ill used if they did not get a new student every day. And they are confident of graduating a class of Ten by the time green peas are ripe.[2]

Emerson's comments suggest that the University had a progressive outlook from the beginning; although the curriculum retained the traditional languages useful for candidates for the ministry, the inclusion of Belles Lettres marked a modernizing trend to the study of literatures written in modern languages, including English. When Joseph Gilmore joined the faculty in 1867, he was appointed professor of rhetoric, logic, and English literature, a title that by the end of the century was simplified to professor of English. Gilmore was one of the first faculty appointments in the country designated specifically as an instructor of English literature, and, in fact, the University of Rochester boasts one of the oldest English departments in an American college or university. Rochester was also one of the earliest institutions to offer courses in American literature where students no doubt discussed Emerson—although, at the time, they could not have known of his early remarks on the University.

Ironically, a full century later Emerson was to resurface at Rochester in a most surprising way—at the center of a national controversy that spotlighted the Department of English's own work in defining the best practices for editing and annotating literary texts for the benefit of students and scholars.

The Emerson Project

Beginning in the 1950s a remarkable group of scholarly editors began to study ways to prepare more accurate and accessible texts of the great American writers. They had begun to realize that many of the accepted versions of literary classics were significantly different from what the authors had originally intended.* Research was also disclosing a large body of interesting, previously unpublished material by writers both obscure and famous. In the process scholars were identifying previously unknown major writers like the Puritan poet Edward Taylor, and they were discovering important unpublished writing by established figures

---

*One Rochester professor, for example, was bemused by a critical essay that tried to make much of a passage in Herman Melville's *White-Jacket* where a sailor is described as falling overboard down past the "soiled fish of the sea." Checking Melville's original edition, it became apparent that Melville was not making theological claims about nature's participation in Adam's sin; he had referred to "coiled fish," and a simple typo in later editions had changed the sense of his image.

such as Emerson, Emily Dickinson, and others. Although an edition of Emerson's journals had appeared in the first decade of this century, for example, few people knew that he had visited the University of Rochester because the passage quoted above had been left out, along with a great deal of other material. Critical, scholarly editing since 1950 has changed the literary canon.

One of the most important and influential literary editorial projects in these years—where Emerson re-emerged as the central historical figure—was headed by Professor William H. Gilman. We now know, for example, what Emerson thought about the infant University of Rochester because Gilman and his colleagues discovered the comments—omitted, along with a great deal of other material, from the earlier edition of the notebooks—while preparing their magisterial edition of Emerson's *Journals and Miscellaneous Notebooks*. In view of Emerson's enormously influential role in the history of American literature and culture, Gilman and his fellow editors thought that his complete text should be available for all readers and not just those scholars able to consult the manuscripts in Harvard's Houghton Library. Their edition in sixteen volumes scrupulously presented all of Emerson's text and identified thousands of now obscure references Emerson made to people he had met, books he had read, and news events of his day. As they examined Emerson's manuscript journals and notebooks, they discovered much more material than they had expected because Emerson often crossed out words, sentences, or whole paragraphs and inserted revised versions of his ideas and observations. Gilman and his colleagues patiently recovered most of these first thoughts, and by printing them side by side with Emerson's revisions they laid out evidence for new and fuller insights into the development of Emerson's mind and writing, giving not just the final thought but also the thinking and re-thinking that led up to it.

Gilman's edition of the *Journals and Miscellaneous Notebooks*, even as it stimulated a renaissance in Emerson studies by presenting new material, showed an Emerson not previously known. Since 1960, when the first volume of this edition appeared, scholars using this material have revealed a more complex, more tormented figure than the cheerful sage many had previously taken him to be, a seriously engaged intellectual as well as a man with a droll Yankee sense of humor. Until Gilman showed passages like Emerson's sly description of the University of Rochester's hopes of graduating a class "by the time green peas are ripe," how many readers knew Emerson was funny? Earlier studies of Emerson had represented him as relatively unconcerned with the abolitionist movement, but Gilman uncovered substantial journal passages where he describes Emerson as being "angry, bitter, ironical" about slavery and holding "this mood for days." Gilman pointed out, "Analogous to Tennyson, Arnold, and other Victorians in England at mid-century who were troubled by the dying of old worlds and the changing into new in religion, science, manners, Emerson in [his] Journals . . . reacts to the Fugitive Slave Law as symbolic evidence that the world he knew and had often heralded seems to be breaking apart."

Scholarly reevaluations of Emerson have in turn led to more popular biographies and studies that have engaged the attention of a wide variety of readers outside of academia. The new appreciation of Emerson this project encouraged has established him as a central figure in the American literary tradition. He is now widely recognized as a writer who influenced Thoreau, Whitman, Emily Dickinson, Frost, and Hart Crane, as well as a thinker who had significant impact on the American pragmatist philosophers Charles Sanders Peirce, William James, and John Dewey.

Controversy and Criticism

Gilman's work as an editor might have been more important for the methods he developed of transcribing and representing manuscript texts than for the text itself, important as that is. Certainly it brought him considerable notoriety when he and other scholarly editors came under attack in the 1960s. Gilman's Emerson edition was part of a larger effort sponsored by the Modern Language Association to provide accurate, authoritative editions of American writers. The Center for Editions of American Authors (CEAA) had established under its auspices standards for authoritative editing of previously published works, but the Emerson *Journals and Miscellaneous Notebooks* project was the first to attempt a large-scale edition of material that was still in manuscript form. Gilman pioneered conventions for representing manuscript texts on a printed page, and his protocols for editing texts and inspecting the edited copy guided later projects that carried the CEAA stamp of approval.

Gilman and his editors decided to try to reproduce the form of the manuscript on the page by using special symbols to indicate where and when words had been struck out, inserted later, written above the line, in the margins, overwritten, and so on. (Simple photographic reproduction of the pages would not have recaptured passages that had been covered over or otherwise obscured.) This produced a text that was heavily marked by up- and downward-pointing arrows, brackets, parentheses, and that included all possible versions of Emerson's entries. The complicated scholarly apparatus that accompanied this edition was not unusual in itself. Editors of previously published texts such as Hawthorne's novels compared each edition printed during the author's lifetime and prepared lengthy lists of variant wording, spelling, punctuation, even hyphenations. In 1967, however, the culture critic Lewis Mumford delivered a broadside against the new editions, focusing on Gilman's Emerson, with an essay in *The New York Review of Books* entitled "Emerson Behind Barbed Wire." In the following year Edmund Wilson renewed the attack with two essays later printed in a pamphlet as *The Fruits of the MLA*. Suddenly Gilman and his colleagues, engaged in what many other literary scholars had regarded as the most conservative kind of work, found themselves in the front lines of a 1960s' culture war.

Wilson may have been frustrated because his pet project of publishing an American equivalent to the French Pléiade editions of classic French writers had gone nowhere, while the National Endowment for the Humanities was funding the CEAA editions, which in his view were unreadable. Mumford and Wilson objected to the scholarly apparatus as unnecessary and discouraging for ordinary readers, but they missed the point. The CEAA scholars had all along envisioned their edition as useful to later editors who could exercise their own judgment about which versions to use in editions prepared for broader, nonscholarly audiences. As Gilman explained, "One of the aims of these editions is total accuracy; it is to guarantee that the work will never have to be done again." In the long run the decisions of Gilman and the other scholars were borne out. When in the 1980s the Library of America began publishing editions of America's greatest writers in Wilson's preferred format of elegantly printed and manageable volumes, it used texts established by the CEAA editors whenever possible. The impact of the critical editing techniques developed at Rochester and elsewhere has gone beyond the preparation of literary texts. The work of Gilman and other textual critics such as Fredson Bowers and Thomas Tanselle has encouraged historians to rethink their own methods of dealing with manuscript documents. Projects such as the ongoing edition of the correspondence of George Washington and the formation of a multidisciplinary Society for Documentary Editing build upon the work of the Emerson editors and other like-minded scholars.

Companion Projects in Textual Editing

Gilman's work on the Emerson journals and notebooks was a high point in the history of modern textual scholarship, but he was not alone in the English Department in producing important and influential editions. His colleague Cyrus Hoy, who taught Shakespearean and Renaissance drama, edited definitive scholarly volumes of the work of Thomas Dekker and Beaumont and Fletcher. Hoy was also the general editor for the most popular reprint series of Jacobean and Stuart plays, providing accurate texts for classroom use across America. Howard Horsford produced a magisterial edition of Herman Melville's journals that has become essential for all Melville biographers. There are, however, other tasks for editors beyond the restoration and presentation of texts, and George Ford, who oversaw the Victorian sections of the *Norton Anthology of English Literature* through five editions, was in a different way perhaps as influential as Gilman. In the last four decades, the *Norton Anthology* has been used in virtually every college in the country in British literature courses. George Ford's selections of authors and texts and his informative introductions and helpful notes have powerfully shaped the ways in which most students in the last forty years have understood British nineteenth-century authors. Modern scholarly editors not only take responsibility for the accuracy and completeness of texts, as Gilman did; they also assume, as Ford did, a responsibility for providing the additional information that readers might need for the fullest appreciation of a text.

Major editing projects continue to flourish in the English department of today. Jarold Ramsey's edition of Native American oral stories
and poems, *Coyote Was Going There*, has been a fundamental text for
the new field of American Indian literature, both for its selections and
for its astute consideration of how to assess written versions of work
originally intended for spoken performance. Frank Shuffelton's edition
of Thomas Jefferson's *Notes on the State of Virginia* is the first important
edition of this text in nearly fifty years.

A much more ambitious project, however, is the Middle English Texts
Series (METS), directed by Russell Peck. This has since 1990 published
thirty-two volumes containing approximately two hundred individual
works. When completed, the series will include about sixty volumes
and will make permanently available to students and scholars reasonably priced, accurate editions of medieval texts. This project has already
revolutionized the teaching of medieval English literature and history
in North America, Europe, and Australia, and is being used in Japan
and Korea as well. The printed versions of the texts are published by
arrangement with Medieval Institute Publications, and the intention is
to keep them in print in perpetuity. Perhaps more important, the Middle
English Texts Series uses electronic media to present and disseminate its
editions. The Medieval Institute at Western Michigan University maintains a METS web site where edited texts are available in electronic
form (www.wmich.edu/medieval/mip/index.html), and the web site has
a mirror facility at the University of Gröningen in the Netherlands.
Electronic publication is the new frontier for editors and for readers as
well, and Russell Peck's METS project has made it possible for a reader
in a remote (but web-connected) location in today's world to have access to a collection of medieval texts far larger than would have been
practically available to a contemporary of Chaucer.

Other scholars are also working to extend the possibilities of this
technology for editors in exciting new directions beyond those of Peck's
editions. Morris Eaves, an authority on the poet William Blake and
editor of *Blake: An Illustrated Quarterly*, published at the University of
Rochester, has been working on hypertext editions of Blake's illumi-
nated books. Blake created a series of books that he printed from etched
plates which contained both poems and his own illustrations. He subsequently hand-colored each copy, often using different colors for dif-

ferent individual copies and amending text and drawings as well; in effect, each individual copy of a Blake book is its own separate edition. Eaves and other Blake scholars are presenting as many possible versions of these texts in the William Blake Archive, a web site containing a hypermedia treasury of related texts, documents, and images (www.iath. virginia.edu/blake). Advanced principles of design allow the Blake Archive to integrate editions, catalogues, databases, and scholarly tools into one electronic source. Students of Blake are able to access a treasury of Blake images and texts that were previously available only in rare book libraries in America and England, and they are able to search both the texts and, for the first time in any medium, the images with tools made available on the site. With the related texts and images now available, scholars and general readers will be able to gain new insights into the work of this sometimes difficult and challenging poet-artist.

The Writers at Rochester

The work of Gilman and his colleagues, as well as that of all of the other editors in the English department, has always looked ultimately toward bringing literature into the lives of ordinary readers. In this they have been at one with the intentions and work of the department's creative writers in working to bring the power and clarity of literature to as many people as possible. The Rochester department has boasted a succession of distinguished poets and novelists, beginning in the post-World War II era just as the Emerson project was taking shape.* The first of these was Hyam Plutzik, who joined the English department in 1945, shortly before Gilman's arrival, and who gained increasing recognition during his years at Rochester. Plutzik was felled by cancer in 1962, but his creative successors have included Pulitzer Prize winners W. D. Snodgrass and Anthony Hecht (who eventually left Rochester to become the national poet laureate), and Joanna Scott, winner of a

---

*The University's earliest important novelist actually was a student in the 1850s, Albion W. Tourgee, who went on to write the best-selling *A Fool's Errand* and *The Invisible Empire*, books critically portraying the reconstruction-era South and the rise of the Ku Klux Klan.

MacArthur "genius" grant and the prestigious Lannan Literary Award. Other successful creative writers have included the novelists Shirley Schoonover, Charles Flowers, Jesse Hill Ford, and Thomas Gavin, as well as the poets Jarold Ramsey, Barbara Jordan, and James Longenbach.

Hecht, Ramsey, and Scott, among others, have directed the program of readings known as the Plutzik Series, endowed in memory of their predecessor, which brings nationally and internationally known writers to read their work before local audiences, and Rush Rhees Library now houses in the Plutzik Center for Contemporary Writing a remarkable collection of first editions of modern poetry. Yet, if these writers are trying to bring their work before contemporary readers, they rely on editors to present it, and in the future their work will need editors to preserve and explain it for generations of readers to come. Gilman's and Ford's editorial projects have changed the way people at the end of the twentieth century look at the traditions of English and American literature, and the projects of Eaves and Peck will invigorate the readers of the twenty-first century. Green peas have ripened 150 times since Emerson visited the brand new institution on Rochester's Main Street, but from writer to editor to reader the circle of language continues.

—Frank Shuffelton

8

Positive Political Theory

"I visualize the growth in political science of a body of theory somewhat similar to . . . the neo-classical theory of value in economics."
—*William H. Riker*

At the close of the twentieth century, the landscape of political science
would be entirely different if not for the symbiotic partnership forged
between William H. Riker and the University of Rochester's political
science department. Riker was a theoretical visionary and masterful in-
stitution-builder who helped transform political science from its 1950s'
state of methodological confusion to a unified science of politics on an
equal footing with physics and economics. By the 1990s, Riker's "positive political theory" would be recognized as a mainstream research platform with adherents at all major American political science departments.

Also referred to as "rational choice theory," positive political theory
has two major theoretical components. One is that political science
should be comprised of statements deduced from basic principles that
are subject to empirical test. The second is that individual decision-
making, premised on rational self-interest, is regarded as the source of
collective political outcomes. The successful promulgation of this "Rochester" theory can be attributed to Riker's dynamism as an intellectual
and institutional leader, to the University of Rochester's supportive administration, and to Riker's capable and enthusiastic colleagues and students. In particular, Richard F. Fenno, Jr., who was Riker's colleague
throughout the latter's tenure at the University, played a critical role in
training new scholars and demonstrating the compatibility of positive
political theory with groundbreaking empirical work.

Riker's Intellectual Formation

William Riker earned his Ph.D. at Harvard University's government department in 1948, having written his dissertation on the Council of Industrial Organizations. Yet he remained dissatisfied with the then-dominant case-study approach to political science exemplified by his own dissertation. Throughout his career he would oppose the case-study approach which, in his view, was ill-conceived because it sought idiosyncratic explanations of individual events instead of seeking to identify general patterns with universal validity. Riker also questioned the contemporary tendency to view political science as a normative enterprise geared toward evaluating political events by using assumed values and ethical postulates. The method of positive political theory which Riker would strive to construct throughout the 1950s countered the tendency to concentrate on individual events with a commitment to find universally applicable laws, and countered the tendency toward value analysis with a commitment to draw a sharp distinction between facts and values.

On the faculty of Lawrence College, Appleton, Wisconsin, where he headed the small political science department throughout the 1950s, Riker was captivated by a new style of literature flowing from the RAND Corporation and elsewhere. This literature attempted to tackle the analysis of political problems, such as elections, using quantitative and logical means. L.S. Shapley and Martin Shubik constructed a mathematical index for political power pertaining to legislative votes. Kenneth Arrow, with his famous "Impossibility Theorem," demonstrated the logical impossibility of a general method of aggregating voters' preferences in a fair and representative way. Arrow's student, Anthony Downs, developed an "economic theory of democracy" which argued that certain features of election outcomes can be predicted by analyzing the self-interested actions of constituents.

Riker also read John von Neumann and Oskar Morgenstern's *Theory of Games and Economic Behavior* (1944). With half a century of hindsight, the significance of game theory for political science and economics is startlingly evident, but that was not the case in the late 1940s. Though RAND mathematicians did investigate the applicability of game theory to problems of nuclear strategy in the 1950s, Riker was the first non-RAND theorist to recognize its potential to analyze political interactions. Similarly, Riker recognized the value of the pioneering work by Duncan Black, who used formal analysis to show that the difficulty of fairly aggregating voters' preferences can be overcome if they are patterned in certain ways.

Riker was the first professor of political science to grasp the promise of these innovative approaches for designing a new methodology for political science, and he was quick to introduce these approaches into his curricula at Lawrence College. He wrote articles drawing on Shapley and Shubik's power index, and he explored whether Arrow's theorem about elections was relevant to actual voting outcomes. Interested in a broadly cohesive basis for a science of politics, he wrote two papers for philosophy journals which directly challenged the anecdotal, narrative history approach still widely used by political scientists in the 1950s. His intellectual curiosity about game theory spread among his undergraduate students, who participated in experiments he initiated on political coalition-building. Riker used game theory to predict how students would form coalitions in bargaining situations, and he later incorporated these experimental results in his book on political coalitions.

By 1959 Riker's theoretical synthesis for a new science of politics was complete. In an application to the Center for Advanced Study in the Behavioral Sciences (CASBS) at Stanford University, he described his new field of study as "formal, positive, political theory," and explained, "By formal, I mean the expression of the theory in algebraic rather than verbal symbols. By positive, I mean the expression of descriptive rather than normative propositions." He then elaborated his plan for his discipline:

> I visualize the growth in political science of a body of theory somewhat similar to . . . the neo-classical theory of value in economics. It seems to me that a number of propositions from the mathematical theory of games can perhaps be woven into a theory of politics. Hence, my main interest at present is attempting to use game theory for the construction of political theory.

As a fellow at the Center in 1960-61, Riker articulated the basic tenets of his freshly minted positive political theory in *The Theory of*

Political Coalitions (1962), which became one of the canonical texts in the burgeoning rational choice tradition.* In the book, Riker used game theory to build predictive models of political interactions, and he was the first political scientist to integrate logical analysis into his methodological tool box. Drawing on von Neumann and Morgenstern's theory of games, Riker modeled political action based on the premise of agents' self-interested, strategic decision-making. Here Riker paralleled the well-known models of economists who demonstrated the balance struck between supply and demand under the assumption that individuals seek to maximize their satisfaction through calculated consumption. However, in Riker's theory of politics, agents are rationally strategic actors who seek to "win" through deliberate decisions, and the task for political scientists is to predict the outcome of agents' collective interactions. Riker distinguished his work from normative political science, from the standard political science "wisdom literature" built on anecdote, and from the growing behaviorist movement, which used statistical methods but did not treat political actors as rational agents seeking to achieve goals.

Building a Department

Had Riker either continued his career at Lawrence College or accepted a post at one of the country's leading political science departments in the 1960s, it is almost certain that he would not have had such widespread impact on the entire discipline of political science. His methodological vision and institution-building genius fit naturally with the University of Rochester's commitment to building strong, quantitatively oriented social science departments. This union elevated "Rochester" from a little-known name in the American social sciences to a nationally recognized school of thought by the final decade of the century.

Before setting forth to the CASBS, William Riker had come to the attention of University of Rochester administrators who were keen to

---

*Beside the aforementioned works by von Neumann & Morgenstern, Black, Arrow, and Downs, other classic texts include James Buchanan and Gordon Tullock's *The Calculus of Consent* (1962) and Mançur Olson's *The Logic of Collective Action* (1965).

establish graduate programs in the social sciences, and who had the underlying fiscal resources to proceed on this course. Joseph C. Wilson, chairman of the Board of Trustees, was personally committed to the ideal of science as a means to better human lives, and he was eager that the Rochester social science programs play a role in this undertaking as vital as the already recognized natural science programs. Riker, whose research admirably fit this bill, was hired to establish the graduate program in political science. At about the same time, Lionel McKenzie was brought on board to chair the economics department, and W. Allen Wallis, former dean of the University of Chicago's business school, was selected as the University's president. Like Riker, Wallis and McKenzie were both committed to the project of developing analytic and formal methods in the social sciences.

When Riker arrived at the University of Rochester in the fall of 1963, the small political science department had some very talented faculty members, most notably Richard Fenno. At that time, Fenno was studying appropriations politics in Congress, a project that would soon lead to the publication of *The Power of the Purse*. This book signaled Fenno's emergence as a major student of Congress. In ensuing decades, Fenno would establish himself as the nation's preeminent Congress scholar. In recognition of this achievement, Fenno, like Riker, would be elected president of the American Political Science Association.

Still, the department was small when Riker arrived; there were few faculty and no graduate students, and it was nearly invisible to the rest of the discipline. Riker ambitiously set out to revamp the undergraduate major, and to found a serious graduate program using the blueprint of his positive political theory.

Within weeks of his arrival he proposed an entirely new curriculum of fourteen courses and seminars ranging from "theories of strategy" and "problems in measurement of political events," to "political parties" and "legislative behavior." From the outset, Riker sought to rival the then-nationally recognized programs housed at Yale, Chicago, Northwestern, MIT, and the Michigan Survey Research Center. In his written proposal to the graduate dean, Riker laid out "a two-fold emphasis . . . on 1) objective methods of verifying hypothesis (i.e., 'political behavior'), and 2) positivistic (i.e., non-normative) theories of politics."[1] The new Ph.D. program required students to gain fluency with quantitative methods and formal analysis. In an unprecedented move, Riker persuaded the graduate dean to accept the substitution of statistics for a modern language. Here he distanced his program from others by its emphasis on scientific training in opposition to knowledge of vast bodies of literature.

Faculty recruitment was Riker's next priority. Over the next few years, he brought in Gerald Kramer, Arthur Goldberg, John Mueller, Richard Niemi, Alvin Rabushka, Gordon Black, G. Bingham Powell, and Bruce Bueno de Mesquita to join Riker, Fenno, William Bluhm, and Peter Regenstreif. Similarly, Riker strove to attract both undergraduate and graduate students, and the results of his energies were readily apparent. By the early 1970s, the department was flourishing with over 100 undergraduates and between 25 and 30 graduate students. As of June 1973, it had graduated 26 doctoral students and 49 master's students; it moved up in the American Council of Education ratings from being unrated in 1965 to holding fourteenth place in 1970. In student placement during the 1960-1972 period, Rochester's political science program was second only to Yale's. In the next decade, these students would be appointed to the faculty of numerous institutions with nationally recognized programs, and this first generation of Rochester Ph.D.'s, coming from a then-unknown program, would be crucial in transforming the discipline of political science in the upcoming decades.

The Rochester political science department esprit de corps was aided by the University administration's steadfast commitment to providing the resources Riker required to maintain his faculty and build a flagship program. Other political science departments were quick to notice the marshaling of a leading program at the University of Rochester and mounted efforts to steal away Riker's carefully assembled faculty and, indeed, Riker himself. Despite these constant attempts, only two (Kramer and Goldberg) left the department during Riker's first decade as chair. Throughout this period of early institution-building, Riker remained pointedly focused on his comprehensive program for erecting the study of politics on a new foundation. In a letter to the graduate dean, he observed, "One main reason for this departmental success is, in my opinion, the fact that the department has had a coherent graduate program, centering on the notion of rational choice in political decision-making."[2]

Transforming a Discipline

The decades following Riker's intensive efforts to build the Rochester graduate program in political science were devoted to spreading the rational choice approach to departments across the nation. The steady achievement of professional milestones indicated not only that the Rochester school had matured as a subfield of political science, but that it had secured its legacy within the entire discipline of political science. Riker's personal publication record did much to raise the visibility of the rational choice approach. In 1967, the editor of the leading political science journal, the *American Political Science Review*, wrote Riker, "There is some danger of turning this journal into the 'William H. Riker Review'."[3] In 1972, with his former student Peter Ordeshook, Riker published a textbook which made positive political theory accessible to students beyond Rochester.

Buttressing Riker's personal prominence in the field, Rochester's first generation of graduates built successful careers and thus introduced positive political theory to departments across the nation, establishing strongholds and new outposts at Caltech, Carnegie Mellon, Washington University, Michigan State University, and elsewhere.* By 1986 Morris Fiorina and Kenneth A. Shepsle, who were among Fenno and Riker's first students, had attained appointments at Harvard, affording Riker special pride. In Riker's view, his alma mater had long insisted on perpetuating outmoded, nonscientific approaches to politics, and had at last come around to acknowledge the leading role that positive political science rightfully played.

Meanwhile, benefiting from Rochester's strong pedagogical tradition, a second generation of students graduated to reinforce the numbers who were already practicing in the field, while more were being trained under the auspices of Rochester graduates at other institutions. Like their predecessors, the second generation of Rochester students have become prominent contributors to many subfields in political science.

*The first-generation of students included Peter Ordeshook, Kenneth Shepsle, Barbara Sinclair, Richard McKelvey, John Aldrich, David Rohde, Morris Fiorina, and Keith Kochbiel, among others.

Their impressive scholarly productivity ensured that Riker's pioneering vision would become one of the field's standards.

Riker's rational choice theory of politics gained additional preeminence when first he, and then other Rochesterians including Fenno, Shepsle, Richard McKelvey, and Fiorina, were admitted to the National Academy of Sciences (NAS) in the 1970s and 1980s. Admittance into the NAS was at least in part due to the embrace of natural scientists who deemed that Riker and his contingent had met the dictates of rigorous scientific inquiry. Similarly, the Rochester school's positive study of politics was heralded as one of the discipline's standards when Riker was selected to serve as the president of the American Political Science Association in 1983. His presidency signaled that all political scientists, whether sympathetic or not to the Rochester approach, had to acknowledge that positive political theory had forever changed the terrain of political science. In the next decade, all major departments would have faculty who worked within the rational choice research tradition. Today, scholars with Rochester Ph.D.'s can be found on faculties across the nation* where they continue to engage in researching, teaching, and mentoring new generations of students.

A Lasting Legacy

The Rochester positivists have altered the discipline of political science by introducing new methods for studying political events and contributing new insights into political processes. Most important, the Rochesterians focused political analysis on the problem of predicting collective outcomes as a result of individually rational decisions. They have applied this method to political processes (such as elections and the platform formation of political parties), legislative behavior (such as coalition formation and bargaining), public goods (such as the "tragedy

*These faculties include: American, Brigham Young, Brown, Caltech, Carnegie Mellon, Columbia, Dartmouth, Duke, Florida State, Kentucky, Harvard, Miami, Michigan State, Northwestern, Penn State, UCLA, William and Mary, and Yale, among others.

of the commons" and the "free rider" problem), and treaty formation and diplomatic strategy in international relations.

Positive political theorists are at the forefront of studying complex collective action problems that dominate our age. The department continues to support one of the country's outstanding research faculties and remains devoted to training many of the discipline's most influential scholars and teachers. As it enters a new century, the Rochester political science department remains a leading center for the scientific study of politics.

—S. M. Amadae

9

Geriatric Medicine's Coming of Age

"Growing old is not a disease."

—T. Franklin Williams

"Father of Geriatrics"

It may not be literally true that T. Franklin Williams launched the study of geriatrics in the U.S., but it's close. When he arrived in Rochester in 1968 to assume a joint appointment as professor of medicine at the University and medical director of Monroe Community Hospital (MCH), the field didn't exist as a medical specialty in this country. Williams was one of a handful of medical people who happened to be interested in the special problems associated with long-term care, but nationwide, little attention was being given to training doctors in the specifics of health after age sixty-five. Happily, his new position in Rochester provided the opportunity to begin the work of learning what exactly goes on in the aging body and mind, and to apply those findings to real life situations. The job also gave him the jump on the rest of the country. In the 1970s, new demographic studies showed that the fastest growing segment of the population was people over age sixty-five, and interest increased dramatically. Suddenly there was a demand for medical personnel trained in the care of older people. By that time, Williams, Monroe Community Hospital, and the fledgling program at the University of Rochester were already national leaders in this new and increasingly important field, geriatrics.

University-Hospital Marriage

In the late 1950s, Marion Folsom, vice president of Kodak and a former secretary of the Department of Health, Education, and Welfare, headed up an important set of studies in the Rochester area. These studies, undertaken by regional health planning councils, led to a unique collaboration between Monroe County and the University of Rochester that has had resounding results for both institutions, for the community, and for the entire young history of geriatrics and gerontology.

The county hospital, known then as the Monroe County Infirmary, had lost its accreditation because of its inability to attract qualified physicians to its resident staff. Monroe County needed the professional contribution of the University's medical school. In turn, the medical school needed a chronic-care setting to fill long-standing gaps in its educational programs. In an agreement reached in 1968, the University would provide the medical and dental staff for the newly configured Monroe Community Hospital. The county in turn would compensate the University and provide the setting and support for teaching and research in chronic illness, assuring that the quality of staff and program at the hospital were equal to that of the medical school generally. T. F. Williams was hired to run this unique operation. As he later wrote, "This binding commitment to this field and the resulting developments are, I believe, not matched by any other medical school or community."[1]

Williams's long career in geriatrics began serendipitously. Ever since his Harvard medical school days (1946-50) he'd taken an unusually sharp interest in the matter of "whole patient" health care, especially of the chronically ill. From his Johns Hopkins residency, through his tenure as professor of medicine and preventive medicine at the University of North Carolina, his basic scientific interest (and the subjects of his numerous scientific papers) lay in diabetes and other metabolic disorders that afflict older persons, and require long-term treatments. Diabetes, he observed, "is a kind of microcosm of many of the general problems of the chronically ill. While it can be controlled by diet and drugs, it cannot be cured. By exercising constant care in the way they live, diabetics can help themselves lead normal, useful lives within the context of their condition."[2]

To his new position in Rochester, Williams brought two clear principles: 1) in everything, a significant teaching and research component, and 2) the best possible care for the chronically ill in a humane setting. "Few other settings are able to offer the quality of care, or make the study of chronic illness an integral part of their medical school program," Williams said. "That's why I came. I wanted to be part of this new venture." Under his guidance, the new venture moved forward on three fronts: medical research and training; care for both the healthy and ill older population; and complete reunderstanding of the health implications of aging.

Medical Education

Williams's first goal was to introduce future physicians, regardless of specialty, to geriatrics, giving them a base of knowledge and experience in the care of elderly persons. The University of Rochester was one of the first medical schools in the country to establish a geriatric curriculum for residents and fellows. Each of the four medical-school years includes a geriatrics component. Residents in primary care rotate through MCH as well as other geriatric settings in the community, and there are also experiences for residents and fellows in neurology. Geriatric grand rounds and special neurology and psychiatry rounds are supplemented by regular seminars and conferences. The regular geriatric evaluation/consultation clinic includes several medical students, residents, fellows, and graduate nursing students as well as faculty. All these activities emphasize comprehensive interdisciplinary team experience, including nurses, social workers, and physical and occupational therapists, specialties essential to the care of elderly persons with complex medical and psychosocial problems.

In 1981 Williams resigned as director of MCH to co-direct a new Center on Aging in the medical school. The purpose of the center was to carry out gerontological and geriatric research and education, and, in cooperation with community agencies, to develop and model new services to meet the needs of the area's elderly population. No Department of Geriatrics existed at the medical school, then or now. (Nationally, among 126 medical schools, there are only three full-fledged geriatrics departments; the remainder have geriatric programs of varying strength, including geriatric fellowship programs.) While some worry about producing too few doctors who choose geriatrics as a specialty,* this does not greatly bother William J. Hall, current chief of the General Medicine/Geriatrics Unit at the University's Medical Center. He much prefers Williams's "cowbird" strategy of laying gerontological eggs among the other medical disciplines for them to nourish, thereby getting the field heavily involved in the entire medical curriculum. Gerontology by its very nature is interdisciplinary, a meeting point for medical interests of every variety, with the possible exception of pediatrics.

The Geriatric Assessment Program

Williams's first task, when he came to Rochester in 1968, was to put into place the program that most directly addressed the question of quality care in a humane setting. The 1950s' Folsom studies had pointed to a serious misallocation of hospital beds in the country, and anxieties were running high over rising hospital and Medicare costs, shortages of acute-care bed facilities, and overcrowded nursing homes. One of the bed surveys found "well over half of the patients in long-term care institutions . . . to be at inappropriate levels of care, usually at higher levels than were judged to be appropriate."[3] At the same time, the community "needed few if any more acute hospital beds, provided it gave appropriate care to those who needed less costly levels, but were occupying acute hospital beds."

In response to a recommendation in the "Health Care for the Aged" study, Williams and his colleagues established a demonstration Evaluation and Placement (E-P) Unit for diagnosis, evaluation, and placement of chronically ill and elderly persons facing critical questions about their care, with clinical space and specialized consultative and labora-

---

*A RAND Corporation study in 1980 projected a critical need of 7,000-10,300 physicians committed to geriatric practice by 1990. Today there are 3,000-5,000.

tory services set up at Monroe Community Hospital. It was the first such program in the country. The E-P procedure involved teams of internists with special interest and competence in the problems of aging persons, nurses with training in long-term placement for the chronically ill and elderly, medical social workers, and consultants in rehabilitation, psychiatry, and other areas as needed. Between 1970 and 1973 the unit intensively studied 332 patients, all of whom had been referred—by family, family doctor, or others—for placement in nursing homes.

At the conclusion of the first three years, the E-P unit announced that "two thirds of the patients have been recommended for care other than a nursing home, and in most instances at less intensive levels. This includes almost 23 percent who have been enabled to continue in their own homes with some assistance as needed from Organized Home Care, or from visiting or public health nurses." For about a third of the patients a course of active medical treatment or of intensive rehabilitation therapy was recommended and arranged for, which substantially reduced the level of care needed afterwards. The E-P unit also learned that "a large proportion of elderly people sick enough to be considered for placement in long-term institutions are receiving little or no regular medical care." Despite chronic illness, half the patients did not have a regular physician, and more than half required further medical study or treatment before a long-term decision could be made.

Extrapolating from the demonstration study to the county at large, the report also suggested that "yearly savings of at least $2 million might be achieved among Medicaid patients alone, through universal use of the E-P service."

The importance of the E-P Unit and the continuing assessment program that evolved from it can hardly be overstated. It has powered MCH's transformation from frequently disparaged "county home" to a nationally recognized setting for the best possible care and rehabilitation for people with problems associated with aging and chronic illness. It has provided critical training to University of Rochester medical students and faculty and staff as well as to public health officials throughout the area. It has offered clear and basic research opportunities on medical, clinical, social, and psychiatric fronts. It has spurred the development of alternate forms of assisted living and helped to humanize nursing homes and professionalize home-care assistance. It has been an invaluable component of the hospital's medical "outreach" into the community. It has also been a model nationally and for many other communities and settings, including the Geriatric Evaluation and Management (GEM) units that have been made regular inpatient and outpatient services in almost all veteran medical centers.

Nationwide, despite the alarms of the 1970s and 1980s, programs on the Rochester model have done much to stabilize the demand for nursing-home care. Hospitals meanwhile are begging for patients to fill their acute-care beds, and will probably face downsizing in the coming years. But the largest benefit of the geriatric assessment program has come in advancing the scientific and human understanding of the pathologies, the needs, and the health of older persons.

New Understanding of Aging

Since Williams's appointment in 1968, the aging of America has cast quality-of-life issues for older Americans into especially sharp focus.* Medical advances over several decades have not only prolonged life, but have made many of the longer lives healthier. Yet, says Williams, "our perceptions have not kept pace with our advances. We are just beginning to learn what we humans are capable of doing—that we don't have to expect to live the stereotype of aging. . . . We need to seek a better understanding of what we reasonably *can* expect."[4]

A concern for "functionality" of older persons is vintage Williams, who has always been aware of serving two populations of the elderly:

*The average American lifespan today is 76.5 years, a gain of 30 years since 1900.

In 1997 there were 34 million Americans over age 65, nearly 12.7% of the population, an increase of 3 million in seven years. By 2030 the over-65 population will have grown at more than twice the rate of the general population growth.

In 1997 there were also 15.6 million Americans over age 75, up nearly 2 million since 1990, and 3.9 million over age 85, up 850,000. Today there are nearly 5 million 85 and older.

Furthermore, centenarians are the fastest growing segment of the population: about 70,000 now, but projected to reach anywhere between 800,000 and 4.2 million by 2050.

the twenty percent who use eighty percent of the medical services, and the rest who are basically healthy and need only marginal sorts of assistance. One of Williams's signal messages for physicians, researchers, and older persons themselves is, "You can get old and still be healthy."

In 1983, T. F. Williams was appointed director of the National Institute on Aging (NIA) of the National Institutes of Health in Washington, D.C. From then until his retirement in 1991, this gave him a national platform from which to advocate his twin principles of quality care for the chronically ill and strong programs of teaching and research. The first director, Robert Butler, had laid the groundwork with a receptive Congress, and Williams built on that, expanding research activities through a growing budget—from $150 million to over $400 million—with special emphases in four areas: the basic processes of aging, the clinical aspects of aging, Alzheimer's Disease and other dementias, and the behavioral and social aspects of aging. He lobbied against the mandatory retirement age, and for programs designed to improve the quality of life by extending the healthy productive middle years (ages 40-60). Williams was also architect of a major funding organization, the Claude Pepper Center,* through which he helped shape the direction of research nationally by asking of all proposals: "How does it improve and maintain functionality of older people?"

Williams is, himself, an example of a healthy, productive life. After his retirement from the NIA in 1991, he returned to Rochester as professor emeritus of medicine at the University, and as an attending physician at the Monroe Community Hospital, where he continues today seeing patients and taking part in the teaching and research activities. He remains a tireless advocate for his twin principles throughout the community. Among his many honors (and one which he shares with Robert Butler, the first director of NIA, and with three University of Rochester colleagues) is the Gustav O. Lienhard Award for the advancement of health care,† the highest award given by the Institute of Medi-

*Named for Florida congressman and outspoken advocate for older persons, Claude D. Pepper.

†The other University recipients are Loretta Ford, professor and dean emeritus of the School of Nursing (see "The Unification Model of Nursing Education" in this book), Robert Haggerty, professor and former chair of pediatrics, and Ernest Saward, professor of social medicine.

cine of the National Academy of Sciences. In 1995 he was appointed Distinguished Physician at the Canandaigua VA Medical Center. In November, 1999, he and his wife Carter Williams, a geriatric social worker, were honored for their community service to the Lifespan Program and to the Jordan Health Center in Rochester.

Impact

In Rochester, fittingly, Williams's legacy in geriatrics and gerontology is especially strong. The University hosts one of about a dozen Pepper Centers for Older Americans. The University of Rochester Medical Center has identified aging as one of its research priorities for the twenty-first century. Williams's original Center on Aging at the medical school has been succeeded by the Center on Aging and Developmental Biology under Howard Federoff, housed at the new Arthur Kornberg Medical Research Building.

The local impact of course extends beyond the campus and the hospital into neighborhood health and nutrition centers, assisted living centers, professionalized nursing home care, an educated and savvy corps of private-practice physicians, and institutions such as ACCESS (the Monroe County Long-Term Care Program, Inc.), Meals on Wheels, the Wheelchair Mobile Program, and other services. The Center for Lifetime Wellness collaborates with the Pepper Center and MCH in providing educational, assessment, fitness, and rehabilitation services, as well as advocating for the Williams-inspired concept of a healthy old age.

In addition, graduates and fellows of the medical school and the geriatric program at MCH have moved on to play strong roles on the national scene. Several former MCH fellows and junior faculty now head geriatric programs around the country—for example, Mark Williams at the University of Virginia, Knight Steel at Hackensack University Medical Center, Robert Campbell at Cornell, and Mary Tinetti at Yale. John Rowe, a medical school alumnus, former head of geriatrics at Harvard, now president of Mt. Sinai School of Medicine and member of the MacArthur Foundation Research Network, has coauthored a book on successful aging.[5] Former student Thomas Perls directs the New

England Centenarian Study at Harvard, and in a recent book, *Living to 100*,[6] has taken Williams-style advocacy for "functionality" of old-agers into the next century with the news that "good health is the norm for most centenarians."

Horizons

At the University of Rochester, the immediate future for geriatric research will emphasize both of Williams's twin concerns. Scientists will focus on age-related diseases and disorders—Alzheimer's disease and other dementias, depression, osteoporosis, disabling conditions of the heart and pulmonary system, and risk factors for problems with the immune system. More emphasis will go to gender differences and the particular disorders of women, who comprise seventy percent of the population over the age of seventy.

But there will be increasing attention as well to old-age health, to the issues arising from longer lifespans and the possibilities of greater health and fitness—of social and mental "functionality"—into increasingly advancing years. Centenarians, for instance, are the most swiftly growing population segment, forerunners of "the most significant social trend of new millennium: a greatly extended life span for millions of people; the ninth and tenth decades of life filled with opportunity, lucidity, mobility, and good health."[7] Surprisingly, they are on the whole healthier, and in need of less medical upkeep than those in their seventies. They are the hardy ones who've passed the danger years of death by heart disease and stroke, either because they are genetically favored, or because they've managed their middle years healthily. "The older you get, the healthier you've been."[8]

By about 2003 (or earlier) the human genome-mapping project will be complete. Issues of genetic engineering will grow hot, with theoretical possibilities emerging for isolating and replacing genes identified for risk factors—for dementia, for just one example.

More immediately, according to Hall, the legacy of T. Franklin Williams will be best served by continuing the many-fronted efforts to make Rochester one of the healthiest communities to grow old in; to develop clear science and translate it into action; and to provide good systems of health delivery. The University, says Hall, wants to have a leading role in the primary care of older persons—and to extend the concept of improved "functionality" not just to hospitals and nursing homes, but to wherever they live. That, he says, would be the best tribute to the man whom Senior Vice President and Vice Provost for Health Affairs Jay Stein, in the fifth edition of *Internal Medicine*, calls "the father of geriatrics."

—John Blanpied

10

The Unification Model of Nursing Education

"My major strategy, throughout my career, was to improve patient care by improving nurses' education as competent, confident, compassionate practitioners and scholars, skilled in clinical decision-making and in investigating outcomes."

—*Loretta Ford*

When Loretta Ford arrived in Rochester in 1972 to become the first dean of the University of Rochester's new School of Nursing, she already had established a national reputation as a trailblazer. As one of the co-founders of the nurse practitioner movement, the country's first advanced-practice nursing model, Ford was not only changing the way nurses were being educated, she was radically reshaping patterns of patient care delivery.

Over the next quarter of a century, Ford's reputation continued to grow. Rochester proved to be fertile ground for the development of yet another concept—the "Unification Model"—that transformed the entire nursing profession. In November 1999, she was honored as one of nursing's "Living Legends" by the American Academy of Nursing during an award ceremony in Washington, D.C.

Background

To understand the far-reaching influence of Ford's first research contributions, we have only to remember that as late as the 1950s, nurses were seen by many as "doctors' handmaidens." Even such a simple task as taking a patient's blood pressure was often questioned.

As an Air Force nurse during World War II, First Lieutenant Ford knew from experience that her nurse colleagues were capable of doing much more than they were being asked to do. "At 3 A.M. on a military hospital ward, a sudden and unexpected change in a patient's condition could become life-threatening. When there was no doctor around, the nurse on duty had to become the medical decision-maker, the calculated risk-taker," says Ford. "I wanted to change the playing field so that nurses could be decision-makers at any hour of the day or night—and I wanted to be sure they were educated in a way that would enable them to feel and to be competent advanced nursing practitioners."

Ford's vision went against the prevailing model of healthcare, one already centuries old. In that hierarchical model, the physician and surgeon stood firmly at the top of the ladder, with all other healthcare providers stationed on one or another of the ladder's lower rungs. Somewhere near the bottom of the career ladder were nurses; only orderlies and nurses' aides were in more subservient positions. (In her successive roles as nurse's aide, visiting nurse, Air Force nurse, public health nurse, and professor of nursing, Ford knew well the frustrations resulting from the exclusivity of that medically dominated pattern.)

At the University of Colorado, where she was professor and chair of community health nursing during the 1960s, Ford found a physician colleague whose vision matched her own, pediatrician Henry K. Silver. Silver and Ford decided that something should—and could—be done to provide more and better health care for Denver's inner-city and rural children, two of the most medically underserved populations in the community. They decided to design a curriculum that would provide nurses with the advanced skills they would need to offer the children of the community healthcare services that went beyond those that doctors were providing to include parent education, prevention, and health promotion.

The result was the creation of the nation's first nurse practitioner model, one focused initially on pediatric care. In addition to defining the concept and working out the educational requirements for this advanced practice model, Ford and Silver accomplished something more remarkable: They proved the concept's worth to a skeptical medical community—and to their own doubting nursing colleagues.

"We were working at a time when there was a shortage of physicians," says Ford. "But I wasn't educating nurses to save doctors time and effort. I wasn't interested in creating physicians' helpers. I knew that nurses had an important role of their own to play. Patients often tell nurses things they don't tell their doctors, important things that are crucial to their treatment. I wanted nurses to have the educational opportunity to become more professional, more competent to provide healthcare services, and better able to teach people how to live healthier lives."

If nurses' roles were to expand, issues of accountability and legality had to be met. Implementation of the nurse practitioner model in Colorado demonstrated the concept's soundness and gave impetus to its spread across the country. First one, then another, finally all of the nation's nursing schools added the pediatric nurse practitioner course to their curriculum, and the development of other advanced-practice models expanded the concept. Today, 71,000 licensed nurse practitioners are providing healthcare services in clinics, hospitals, physicians' offices, schools, and nursing homes—and that number is growing.

Rochester

In 1972, President Allan Wallis invited Ford to become the first dean of the University of Rochester's new School of Nursing. Wallis, unlike many others in academia, had a comprehensive view of the state of nursing in the United States. In 1966 he had served as president of the National Commission on Nursing Education, and he knew well both the profession's problems and its potential.

At that time, the University of Rochester had two autonomous nursing units, one educational, the other clinical (the latter dedicated to providing care at Strong Memorial Hospital). In a move that promised greater autonomy for the nursing program, Wallis gave Ford a double title: in addition to being dean of the new School of Nursing she was also named director of nursing services at Strong Memorial Hospital, giving nursing its own voice in the cabinet of deans and in the University administration. The fact that the new dean's office was adjacent to those of other leaders of the Medical Center and Strong Memorial Hospital also sent a strong signal from University leadership: Dr. Ford had arrived—and she and her work were to be taken seriously by her medical and administrative colleagues.

Along with the founding of the nursing school came implementation of the University's comprehensive plan for a much-expanded, modernized, and reconfigured Strong Memorial Hospital. Old-fashioned wards were being replaced by new inpatient units, each with a central nursing station surrounded by patient rooms. Nurse faculty and their physician colleagues realized that advances in technology and changes in patient demographics (more acutely ill and older patients) required a new level of professionalism on the part of nurses, and nursing faculty played an important role in planning the new hospital.

Three years before Ford's arrival, in 1969, a study of Strong Memorial's nursing needs was undertaken by the Bates Committee,* an interdisciplinary group convened by medical school Dean J. Lowell Orbison. The report began by noting what hospital administrators knew only too well—a shortage of nurses had repeatedly forced the closing of hospital beds. Nurses, already overburdened, faced increasing challenges. They were providing care for growing numbers of patients with severe illnesses; they were responsible for more tasks delegated by physicians; and they were confronted with sophisticated and complex new technology. To help them meet these challenges, major changes in the nursing curriculum and nursing practice were required.

The Bates Report pinpointed other problems. The traditional hospital system, hierarchical and medically focused, offered little opportunity for staff nurses to advance their education and their skills. Role models were scarce, and there were few, if any, chances for career advancement. In ambulatory care areas, nursing time and effort were often diverted, as nurses were called on to handle administrative and clerical chores, rather than responding to patient needs.

The Bates Report also frankly highlighted the often problematic relationship between physicians and nurses: "In its efforts to participate in this joint venture [providing patient care] nursing has been hampered by its traditionally subordinate relationship with medicine. . . . Physicians tend to be oriented to diagnosis and treatment, nurses to patient

---

*The committee's chair, Barbara Bates, was professor of medicine.

comfort and coping. When methods of achieving these goals conflict, nursing goals are less likely to prevail. Much of the nurse's time is spent on delegated medical tasks—observations, medications, procedures. These tasks are central to the physician's role. Insistent on their receiving top priority, he may fail to recognize their cost in unmet patient needs."

The Unification Model

No one understood the problem better than the nursing school's new dean. And no one had more energy and determination than she—plus the skills required to improve the situation. Using the Bates Committee's suggestions, Ford set about melding the two disparate nursing units into a unity charged with the three-part mission of providing clinical care, education, and research. On the occasion of her installation as dean, Ford called for "a new order for nursing's direction in education, research, and practice, with the University of Rochester School of Nursing in the forefront as a national trend-setter for improving the delivery of health care."

With the "unification model" firmly in mind, Ford gathered the faculty together to refine and energize the new school's educational thrust, one that also would advance clinical care and research. Associate deans for undergraduate and graduate education were appointed, as was an associate dean for practice. Just as each clinical service in the School of Medicine and Dentistry had its own chairman and clinical chief, clinical nursing chiefs were appointed to lead the nursing specialties (medical-surgical, family health, obstetrics-gynecology, pediatrics, community health, psychiatric-mental health, and, later, cancer nursing).

If nurses were to work effectively in their expanded clinical role, Ford knew they must be taught new skills. They would need to know how to collect data through interviews and physical examinations, and how to share that information in an appropriate manner to enable patients to move away from a position of dependency and assume responsibility for their own health.[1] Additional faculty were recruited to lead expanded courses in clinical decision-making, human anatomy and physiology, and in research methods.

The challenges facing the new dean were enormous. There were few doctorally prepared nurses anywhere in the country who could be recruited to take leading roles in designing and developing clinical practices. The nursing chiefs would also need to negotiate ways that the new nursing model could fit effectively into the complex macrosystem that is a medical center. Within the nursing faculty itself, stress developed as academically focused nurses realized that the unification plan required them to leave their "ivory tower" and serve as clinicians in the hurly-burly world of the hospital and its clinics. Conversely, clinical nurses, until then interested only in patient care, were troubled to learn they must return to the classroom to learn the skills they would need to become nurse-educators and nurse-researchers. As an observer noted, the criteria for performance and productivity of each group were foreign to the other, and their expectations and values were often worlds apart.[2] The skills of a diplomat were required to smooth the ruffled waters. The new dean met that challenge, too.

An immediate problem was how to successfully integrate the nursing staff into the "new" hospital, which was projected to open in 1975. Nursing faculty worked with their medical colleagues and hospital and nursing administrators to design an experimental unit in which to test the new roles nurses and support staff would play in the expanded, 804-bed hospital. Criteria were established for evaluating outcomes in terms of patient care, staff responses, and costs. Clinical experience was used to inform both teaching and research—a concept embodied at the heart of the unification model. By the time the expanded Strong Memorial Hospital opened in 1975, the transition—as it related to nursing care of patients—was seamless.

In 1973, a year after Ford's inauguration as dean, she announced a ten-year educational plan for the School of Nursing. The plan projected an educational continuum that progressed from the baccalaureate level to master of science in nursing, to doctor of philosophy in nursing, to postdoctoral study and new opportunities for nurses' continuing education. With the arrival on the Rochester scene of a growing number of nurses with master's degrees, even skeptical physicians began to appreciate that their nurse colleagues could play an important role in clinical decision-making and scientific inquiry.

By 1979 the unification model had proved its success in Rochester and was attracting national attention. Ford became the movement's dynamic spokesperson. She traveled widely, explaining "the Rochester model," as it came to be called, to leaders at academic centers across the country. Far from being simply "doctors' handmaidens," she told them, nurses could learn to become expert clinicians, educators, and researchers. By qualifying to develop their own practices, nurses would be advancing the quality of their patients' care.

To the nurses of the nation, many of them stuck in what often seemed ill-appreciated, poorly-paid, and dead-end jobs, Ford's message was electric. Nursing, she told them, now had its own career ladder, paralleling that of a physician. Those who accepted its challenges could expect rewards, both financial and in professional satisfaction. As for the latter, in learning to build research into their practices, nurses could contribute to the knowledge base and help reshape the way heath care was provided in America. As nurse-educators mentoring young colleagues, they would help ensure continual improvements in patient care. Each step a nurse took toward her own advancement helped improve the public perception of nursing's value. If nurses would accept the unification principle, the whole profession would flourish.

Soon, "unification" was the new byword at schools of nursing in California, Florida, Utah, and Yale/New Haven, and nurses across the country were going back to the classroom to learn how to integrate clinical care, nursing education, and research. Conversely, faculty were spending time in the clinical area, forging new clinical skills and innovations in the field.

Ford's message was heard not just by nurses, but by those in seats of power. During the 1970s, the School of Nursing at the University of Rochester received funding to establish a summer program during which nursing faculty from other schools would learn how to incorporate unification principles into their own curricula. During the 1970s and 1980s, the Robert Wood Johnson Foundation chose Rochester as the site for two innovative programs—a year-long primary care nurse faculty fellowship program and a two-year nurse faculty research scholars program. Both programs brought faculty from across the country to Rochester to learn how to provide research-based primary care and how to educate students in the model.

As a healthcare statesman and a versatile administrator, as well as a nursing legend, Loretta Ford has garnered many honors, including four honorary doctorates. Key among her other awards are the Gustav O. Lienhard Award* for the advancement of health care, the highest honor given by the Institute of Medicine of the National Academy of Sciences; the Edward Mott Moore Award, presented by the Monroe County Medical Society; and New York State's Governor's Award presented for notable achievement to women in medicine, science, and nursing.

—Nancy Bolger

*Ford is one of four University of Rochester faculty members to have received this highly prestigious award (more than any other institution). The others are T. Franklin Williams, professor of medicine emeritus (see "Geriatric Medicine's Coming of Age" in this book), Robert Haggerty, professor and former chair of pediatrics, and Ernest Saward, professor of social medicine.

11

Quest for Fusion: Laser Energetics

"The holy grail is [fusion] ignition in the laboratory. We have a chance to get very close to it here."
— *Robert L. McCrory*[1]

Has there ever been a more wondrously revolutionary technology than the laser? Just four decades old, starting from Ted Maimon's first palm-sized single-pulse ruby-crystal model in 1960, the machine is now ubiquitous. It underlies enormous military-industrial enterprises; it scans your groceries at checkout. It's bigger than a football field; it's as small as a grain of salt. It's at the heart of the quest for cheap, clean, and abundant energy, and for smarter nuclear weapons. It sits on lab desks supercooling individual atoms; it inscribes a million commands on an integrated computer chip; it cuts out your cataracts. It is essential in medical technology, computers, fiber optic communications, industrial machinery, and in commercial products from bandsaws to CD players to automobiles.

Laser: Light Amplification by Stimulated Emission of Radiation. A coherent and highly directional radiation source. A pulse that couldn't hurt a baby's skin from six inches away; but, amplified to forty-five thousand joules, that heats a hydrogen-filled pellet to temperatures hotter than the core of the sun, crushes the atoms, and causes a tiny thermonuclear explosion.

Which is exactly what Omega does. The Omega laser is the elegant behemoth at the University of Rochester's Laboratory for Laser Energetics (LLE) on East River Road. Designed in the early 1970s, on-line with twenty-four beams by 1980, it made upwards of twenty-thousand shots before retiring in 1991 for a $61 million "upgrade." In 1995 Omega reopened with sixty beams. It is now the world's largest ultraviolet laser, surpassing in size, intensity, power, and precision even its erstwhile cousin, Nova (now dismantled, as the National Ignition Facility is constructed), at the Lawrence Livermore National Laboratory (LLNL) in California.

Rochester's Laboratory for Laser Energetics

The LLE is a unique kind of national laboratory in several respects. It is supported through an unusual combination of government, industry, and university funding. It also houses the only major inertial confinement fusion laser at any university, and therefore carries an educational mission as well as that of research and development. Its staff numbers more than three-hundred persons. Its faculty of forty-three straddles six different departments; seventy-six graduate students study there, one hundred work there. It has helped produce 128 Ph.D.'s.

From its beginnings, the laboratory's mission has been basic research in a university setting. LLE workers in the Optics and Imaging Sciences Group and the Optics Manufacturing Group have opened vistas for the entire world in holography research, medical spectroscopy, and the development of advanced laser-optical products. The lab has also generated much excitement for its discoveries in the ultrafast sciences division where experiments and devices have actually gotten down to the femtosecond (one-quadrillionth of a second) level. The short-pulse-laser technology underlying the work of the 1999 Nobelist in Chemistry, Ahmed Zewail, for instance, was developed at Rochester.

The LLE is registered as a National Laser Users' Facility, open to qualified researchers from all over the world in a variety of fields. This facility has funded more than sixty graduate students from universities other than the University of Rochester. And in 1997, the lab won a five-year $143.5 million grant from the Department of Energy to participate in the nuclear weapons' Stockpile Stewardship Program. (While no actual weapons testing is done at the lab and no plutonium or highly enriched uranium is on the premises, the program helps ensure that the country will not have to perform actual tests.) Achieving fusion energy, the laboratory's main pursuit, is a long-term goal, and more expensive than industry alone is willing to fund. Meanwhile, the Stockpile Stewardship Program guarantees an "immediate customer" for the lab, and provides expanded opportunities for research into light-matter interactions.

Nevertheless, LLE's primary role is its part in the national quest for a sustained thermonuclear fusion reaction, and to that end it is one of four "partners" directly involved in support of the National Ignition Facility scheduled to open in Livermore, California, around 2006. Achieving ignition is the first, crucial step in a worldwide energy revolution; it is what Robert L. McCrory (professor of mechanical and aerospace sciences, professor of physics, and director of the Laboratory) calls the "holy grail" of fusion researchers.

Early Days

In the early 1960s, when the new solid-state laser was coming on the scene as a tool, the University of Rochester was right there with it, thanks partly to the flourishing Institute of Optics. A young plasma physicist, Moshe Lubin, came to the University in 1964-65 to work on lasers. Specifically, he was attracted to the tool for its ability to focus high-intensity radiation to a very small spot, to produce superhot gases (or plasmas) and trap them in magnetic confinement coils called toroidal devices. His objective was basic study—how to fill the device, how fast, how hot, how to hold the plasma—but "within the United States, those were probably the first applications of laser plasmas to the thermonuclear fusion program."[2]

At the Lawrence Livermore lab in California some scientists were examining the same laser principles to see if they could find possibilities for compression and ignition of thermonuclear reactions. Crosscurrents of exchange among the Livermore/government/Rochester researchers in the mid-1960s eventually brought everyone to the same conclusion— that lasers, if timed and focused correctly, could probably generate states of matter that would then lead to a self-igniting inertial confinement fusion reaction.

In 1970, Rochester's Laboratory for Laser Energetics was chartered with the mission "to study the interaction of intense radiation with matter." Rochester's self-chosen role in the national fusion program was to provide some of the early scaling experiments, and to understand the physics on a very basic level. The lab was to be interdisciplinary right from the start, residing within the College of Engineering and Applied Science, with Lubin reporting directly to the University president. The idea was to create a research facility that would have an impact on the field. "There was an important role for us as an Institute of Optics and as a College of Engineering and as one of the main players in the laser fusion business."

By 1972-73 Lubin succeeded in attracting two big industrial participants, Exxon and General Electric, to join with the University and government partners in what they called the Laser Fusion Feasibility Project. Others came in later.* The basic-physics concepts for the Omega laser were laid out in 1974-75, in an atmosphere of unusual independence from government control, thanks to the advanced model of industrial/government/university collaboration engineered by Lubin. As a result of its industrial involvement, Lubin stated, "Rochester started to become the place you went to when you wanted to hear a slightly different perspective on the nuclear energy business." The program thrived, drawing scientists from many different fields—optics, physics, mechanical engineering. "It became one of the three or four programs sitting around the table internationally."[3]

Omega

In 1960, Ted Maimon surrounded a ruby crystal with a flash tube which stimulated the atoms inside the crystal. The crystal was about two inches long, its ends coated with silver mirrors aligned to produce one frequency of bouncing light. One of the mirrors was thinner than the other, to allow some of the light to escape as a beam. That was the first solid-state laser, and all since then work on the same principles. All have 1) a gain medium (usually glass or crystal) that amplifies the light passing through it; 2) an energy pump; and 3) at least two mirrors.

Omega consists of over half a million parts, including mirrors and lenses, optical mounts, amplifiers and filters, twelve-hundred mini-computers

*E.g., Northeast Utilities, New York State Energy Research and Development Authority, Empire State Electric Energy Research Corporation, Texaco, Eastman Kodak, SOHIO, S. C. Edison, and Ontario Hydro.

for aiming the beams, and six million dollars worth of neodymium(Nd)-doped laser-glass beamlines. Since all its massive power would be useless but for the immense precision of its optical components, it is housed inside a football-field-sized "clean room." Adjacent to the great room sits the oscillator that generates the original pulse of light. At one end of the room is the eleven-foot target chamber. And in the center of that is the target itself, a thinly encased pellet of fuel the size of a grain of salt, suspended by filaments of spiderweb. The fuel is a "DT" mixture consisting of two isotopes of the hydrogen atom—the single-neutron deuterium and the double-neutron radioactive tritium.

Twenty-two hundred capacitors under the floor are filled with electrical energy before each laser shot. The energy is released all at once to power 6,696 flashlamps. These pump 215 units that amplify the foot-long laser pulse as it passes through the array of Nd glass beamlines. During this five-hundred-foot speed-of-light journey the pulse is split into sixty beams, each amplified, widened to several inches, filtered, and then converted into ultraviolet light. Each beam is then directed by mirrors and focused by lenses to the target. Because each location is precise within ten microns (a tenth the width of a human hair), the pellet is irradiated uniformly if each individual beam is also of high quality.* The beams have arrived pumped up to forty-five thousand joules. In less than a billionth of a second Omega unleashes sixty terawatts of energy, nearly one-hundred times the peak power of the entire U.S. power grid.

Omega's sixty-beam assault violently vaporizes the pellet's thin outer shell, blasting it outward at speeds greater than six hundred miles a second. The result is an equal-and-opposite inward reaction—an implosion—which compresses the fuel at speeds of around three hundred miles per second to a density two hundred times that of the DT's liquid state, heating it to temperatures required to fuse the hydrogen nuclei: one hundred million degrees C, much hotter even than the photosphere of the sun. The resulting burn throws off newly created helium atoms

---

*In a phase-conversion process known as smoothing by spectral dispersion (SSD), each beam is actually broken down into many thousand beamlets. SSD creates a very uniform spatial profile (averaged over time) for each beam, thus assuring the high quality of individual beams required for uniform illumination.

(alpha particles) that slow down in the dense fuel and heat the plasma further, thereby creating a thermonuclear detonation wave. Energetic neutrons are also released by the thermonuclear fusion reaction, and these are the basis for potentially usable energy. In a fusion reactor, the energy released per event is equivalent in strength to about one quarter ton of TNT.[4]

This is the same kind of reaction that powers the sun and the other stars. In the sun gravity is sufficient to confine the superheated plasma and so sustain the reaction for billions of years. On earth, the awesome power of the hydrogen bomb relies on the fusion reaction in an uncontrolled event triggered by an atomic explosion. The laboratory ambitions may be smaller, but in some ways are more fiendishly complex just because of the scale. How do you confine such stuff at such temperatures and densities for a long enough time to yield a net gain of useable energy?

Fusion Energy

Omega's answer is inertial confinement fusion (ICF), the process described above. It is one of the two main branches of the decades-long quest for fusion power, the other being magnetic confinement fusion.* It's easy to understand the lure of the quest. Unlike nuclear *fission*, fusion leaves little radioactive waste. Unlike coal, it produces none of the carbon dioxide pollution that may be leading to disastrous global warming. Unlike oil, the world supplies of which may be good for fewer than a hundred more years, fusion energy is virtually endless.† What's more, the fuel is cheap—deuterium is extractable from water; tritium can be

---

*ICF is also the approach taken by NIF and its other partners. In Magnetic Confinement Fusion, the plasma is compressed in a toroidal "bottle" of interlaced electromagnetic fields. It's a technically viable rival to ICF, but lacks ICF's pertinence to nuclear-weapons research.

†Compared with even the most optimistic estimates of total earth energy supplies, "fission with breeding represents a 'nearly inexhaustible' energy resource. Fusion has an energy potential of 10,000,000 times that of fission with breeding" (Robert L. McCrory, Jr., "Energy Supply and Demand in the Twenty-First Century," *Journal of Fusion Energy* 8 [1989]: 132).

manufactured from lithium that can be placed in the fusion reactor walls.*

The bright promise of the quest for fusion has prompted substantial funding by the government over the last few decades, both in universities like Rochester, Princeton, and others, and in the national labs of Livermore, Los Alamos, and Sandia. But the trail has also been littered with unforeseen frustrations and setbacks. The goal is simple enough: to get more energy out of the thermonuclear reaction than the energy it takes to make it happen, but it hasn't happened yet. To date, Omega holds the world-record for the energy yield from a glass laser-powered fusion reaction—one hundred trillion neutrons—but that's only *one percent* of the energy the laser requires to start the reaction.

The first huge hurdle to be overcome is achieving ignition—propagating the chain reaction. This may take ten to one-hundred times the energy input currently possible. Another basic problem is controlling the rate of the reaction. Right now, Omega is unique in its ability to manage up to one shot per hour. But any useful inertial fusion energy (IFE) setup would have to handle rates of six-to-ten *per second*. The laser itself would have to be far more efficient than a flashlamp-pumped glass laser like Omega. Diode lasers have the required repetition rate, but the large energies and the cost presently prohibits their use. (Efficient fusion laser development is a research area in itself.) Beyond these challenges are questions of harnessing the energy, protecting the components from the radiating debris, and a hundred others scarcely even imagined. "We've pushed right to the edge of the known," McCrory says. "Everything we do in fusion now is something new."

Despite the immense challenges, no one doubts that hydrogen fusion itself is possible, since it's been demonstrated so resoundingly in the laboratory. And though no one can be certain that ICF will be the route to the holy grail of ignition, some very large national stakes have been laid on that outcome. Labs all over the country, Rochester's in-

---

*Lithium can be used as the means of capturing the energy of the fusion reaction, which it does by moderating the neutron given off in the reaction. Since it is included in many fusion reactor designs, it could also be used to generate tritium, and so it would be performing a dual mission in the reactor.

cluded, are working on various parts of the enormous project. The centerpiece, the National Ignition Facility, is costing upward of $1.2 billion. When it goes on-line it will dwarf Omega and all other ICF facilities in the world, with 192 laser beams capable of generating 1.8 million joules of energy, forty times that of Omega. Laser scientists confidently expect ignition to occur there. Only then will the design of the first actual fusion energy plant begin (although many conceptual designs have been in existence for some time). But we are long past expecting rapid mastering of star-power. At least thirty to fifty years will still pass before commercial fusion operations can be mounted.

Breakthroughs

Bluelight

Robert McCrory arrived from Los Alamos in 1976, attracted by the Rochester lab's growing reputation for cutting-edge work with—by the standards of Livermore, Los Alamos, and Sandia—small resources. He headed up the Theory Group which led the effort (with laboratory scientists Stephen Craxton, Steve Jacobs, Terrance Kessler, John Soures, Wolf Seka, and others) for the lab's most dramatic and far-reaching triumph of its early days: the invention of the "frequency tripling conversion system." This was the means of converting infrared to ultraviolet light, whose shorter waves are absorbed by the fuel pellet much more efficiently than the longer red waves then used by all the national labs. (Anyone could do the conversion, McCrory says now; but achieving an 80 percent efficiency at fusion power irradiance levels was the real challenge.) In 1979-80 the lab made good on the theory by means of the first actual red-to-blue conversion in Omega. Six beams were converted in 1983, all twenty-four by 1985, and just one month later, workers carried out the world's first full-blown successful laser shot using ultraviolet light, and broke the world's record for the greatest-ever energy yield: 165 billion usable-energy neutrons from four trillion watts consumed. The feat put Rochester on the map as a major international player in fusion research. Since then, every other ICF lab in the world has followed Rochester's lead in converting to ultraviolet light.

SSD

In 1988, after the LLE compressed a fuel pellet to two-hundred times the DT's liquid state—the highest density ever directly measured in the laboratory—McCrory began to urge the Department of Energy to support an "upgrade" to Omega.[5] The case was essentially clinched by another breakthrough in 1989, a phase-conversion operation called Smoothing by Spectral Dispersion (SSD). A target not uniformly radiated will develop "hot spots" which make fusion impossible by bleeding off the compressive energy. With SSD, Rochester workers developed a way to vary the wavelengths of the hundreds of thousands of converging beamlets just enough to "smooth" or cover the pellet with an almost perfectly uniform illumination pattern.

Direct drive

This technique in turn paved the way for the use of Omega's "direct drive" mode to do the basic ignition scaling experiments crucial to the national fusion effort. Direct drive—Omega's specialty—refers to Omega's way of bombarding the target pellet directly with its laser beams. Indirect drive, the method used at Livermore and Los Alamos and just about every other lab, encloses the fuel pellet inside a gold or steel housing called a hohlraum, inside which the laser beams are converted to X-rays. So far, this indirect method has been preferable because it has been considered much easier to achieve the crucial uniformity of compression required. But the direct-drive approach, while requiring more finesse, has the decided advantage of being more efficient. In direct drive, between seventy and eighty percent of the laser energy is absorbed by the pellet, compared to about ten percent with indirect drive. Direct drive therefore uses far less energy, and so far, at least, it holds the record for the greatest yield.

Although Omega can also perform precision indirect-drive experiments to complement those at the national labs and so help physicists to understand the basic physics behind both methods, McCrory confidently expects it to show the efficacy of the direct-drive approach. NIF meanwhile is hedging its bets, prepared to modify its driver to the Rochester model if and when its superiority is proven.

And what of Omega's future once NIF is up and running? Again, McCrory is confident. LLE people will handle the conversion to direct-drive at NIF. They will build the instruments, do the diagnostics, carry out much of the work. There will be immense challenges well into the twenty-first century in the fusion-energy quest, and no one doubts that the Rochester laser lab will be making crucial contributions.

—John Blanpied

12

Agency Theory

"The problem of inducing an agent to act as if he were maximizing the principals' welfare is quite general. It exists in all organizations and in all cooperative efforts."
—*William H. Meckling and Michael C. Jensen*

When University President W. Allen Wallis decided to build a first-class business school in Rochester, he recruited as dean a visionary who believed in the power of economics to solve a host of problems.

Arriving in Rochester in 1964, William H. Meckling was already a noted economist best known for his analysis and leadership in support of an all-volunteer U.S. armed service. During his nineteen-year tenure as business dean, Meckling not only met Wallis's challenge but also played a crucial role—as both participant and catalyst—in revolutionizing the kinds of questions asked in business. In the process, the school he built made enormous contributions in the critical areas of finance, accounting, and organizational theory. Those contributions, in turn, shaped the research agenda of a generation of business scholars, influenced teaching in graduate business programs across the country, and forever changed how countless companies and executives in the U.S. and abroad conduct business.

Meckling set the stage for such remarkable achievements by adopting an economics-based approach to problem-solving and by committing the school to the scientific method of research. He required that research have an empirical orientation, and pushed his colleagues to see if their data explained the real world. He also turned a small undergraduate and evening business school into a nationally ranked program by recruiting a first-rate faculty and transforming the business college into a graduate business school. A consummate scholar, he constantly probed and questioned, encouraging his faculty to produce their best work, while requiring the same level of scholarship in his own research.

An influential paper, written by Meckling and a young faculty member, Michael C. Jensen, first put a national spotlight on what today is known as the William E. Simon Graduate School of Business Administration. When "Theory of the Firm: Managerial Behavior, Agency Costs, and Ownership Structure" appeared in 1976 in the *Journal of Financial Economics*, the leading finance journal of the day, it quickly brought its authors widespread recognition and formal honors. Within eight years of its publication it was identified as one of the most cited items in its field and named a Citation Classic by the Institute for Scientific Information. The paper stimulated rich research by scholars at Simon and countless other institutions because it set forth a theory that has served as a useful tool for analyzing and explaining the contractual arrangements of organizations, especially the complex arrangements of the modern corporation.

Agency Theory

In 1973 Meckling and Jensen (now a member of the Harvard Business School faculty) began conducting research in Rochester on agency theory. They examined the agency relationship, and the costs of such relationships. Simply put, the agency relationship is the contract under which one or more persons—the principal(s)—engage another person—the agent—to perform some service on their behalf, which then requires that the principals delegate some decision-making authority to the agent. The costs associated with such relationships, such as monitoring and bonding expenditures, are referred to as agency costs.

Where Meckling and Jensen differed from other scholars was in how they approached their subject. Most of the literature of the day focused on the way parties should structure contractual relations, including compensation incentives, in order to induce an agent to maximize the principal's welfare. Meckling and Jensen assumed individuals solve these problems. They focused on understanding each party's incentives that, in turn, shape the structure of the contracts people form. They also

viewed the participants in any agency relationship as self-interested individuals, which opened up new insights into the behavior of managers
and firms.

While they limited the discussion in their paper to the contractual
arrangements between the owners and top managers of corporations,
they pointed out the broad nature of the "agency problem" and recognized that their theory could be applied in many different kinds of situations.

"The problem of inducing an 'agent' to behave as if he were maximizing the 'principal's' welfare is quite general," they said. "It exists in
all organizations and in all cooperative efforts—at every level of management in firms, in universities, in mutual companies, in cooperatives,
in governmental authorities and bureaus, in unions, and in relationships normally classified as agency relationships such as those common
in the performing arts and the market for real estate. The development
of theories to explain the form which agency costs take in each of these
situations . . . will lead to a rich theory of organizations which is now
lacking in economics and the social sciences generally."

Powerful Theory

Although Meckling and Jensen's published work contained concepts
that existed in the literature of the time, it put those concepts together
in a "quite powerful way," explains Clifford W. Smith Jr., a business
faculty member since 1974 whose own work in corporate financial policy,
derivative securities, and other areas benefited from their work in agency
theory.

Although the famous paper dealt primarily with the economics of
the contracting process, it provided a general framework for thinking
about how two individuals with overlapping but not coincident interests agree to get together and structure a contract that guides them in
their work together, as employer and employee, to achieve a set of goals.
The paper also extended this discussion to a corporate finance problem:
How do you structure ownership of a firm?

According to Meckling and Jensen, "contractual relations are the essence of the firm." They argued that the firm serves as a focus for a

complex process in which the conflicting objectives of individuals (e.g., owners of labor, material, capital inputs) are brought into "equilibrium" within a framework of contractual relations in order to maximize profits. Creating that equilibrium, they said, involves real costs to the parties involved.

It is virtually impossible, they maintained, for any agency relationship not to have agency costs. Principals, for example, have monitoring expenditures. These might include observing the behavior of the agent, and/or efforts to control that behavior in the form of budget restrictions, compensation policies, operating rules, etc. The agent may have to expend resources (bonding costs) to guarantee that he won't take actions harmful to the principal or provide compensation if he does. In addition, Meckling and Jensen argued, since every agent has his own self-interests and no agent acts perfectly on behalf of the principal, "there will be some divergence between the agent's decisions and those decisions which would maximize the welfare of the principal. This is the principal's 'residual loss' and it, too, is a cost of the agency relationship."

Opening New Doors

In the agency paper Meckling and Jensen posed questions—at the time unanswered in the existing scholarly literature—regarding the very existence of firms. Why, they asked, "given the existence of positive costs of the agency relationship, do we find the usual corporate form of organization with widely diffuse ownership so widely prevalent? If one takes seriously much of the literature regarding the 'discretionary' power held by managers of large corporations, it is difficult to understand the historical fact of enormous growth in equity in such organizations, not only in the United States, but throughout the world."

Why, they added, would millions of individuals be willing to turn over much of their wealth to organizations run by managers who have so little interest in their welfare? Even more remarkable, why would they be willing to make such commitments on the anticipation that these same managers will operate the firm so that there will be earnings which accrue to the stockholders?

Meckling and Jensen established the major foundation for their own theory of the firm by analyzing the agency costs of equity and debt. They then synthesized their basic concepts into a theory of the corporate ownership structure, taking into account the trade-offs available using inside and outside equity and debt. They discussed extensions of their analysis, detailed some qualifications, and noted areas where more research was needed.

The overall impact was remarkable. Together, Meckling and Jensen, through their research on agency theory, opened the door to the development of a rich analysis to explain corporate decision-making. That analysis, ongoing since the 1970s, has been conducted by many different business scholars in many parts of the world.*

In addition, by the time the agency paper was published in 1976 Meckling and Jensen were already teaching an innovative course in the business school on the economics of organizations. Their course, based on the writings of many people, explored how economics can help managers better structure and manage firms.

Through their research and teaching, the two Rochester scholars stimulated great interest in the economics of organizations and encouraged related research in economics, finance, and accounting. In fact, Meckling's own faculty included two scholars, Jerold L. Zimmerman and Ross L. Watts, who applied agency theory in their own accounting studies. Their subsequent research made them, in turn, catalysts for moving the accounting discipline to an entirely new level.†

Over the years, Meckling and Jensen remained at the forefront in the field of organizations. Of particular note, they impressed upon many economic scholars, and the Simon faculty in particular, the importance of three critical aspects of corporate organizations:

---

*Some of the topics explored by Rochester's own graduate business faculty include bond covenants; corporate debt; executive compensation; performance evaluation; cost allocations; franchise contracts; corporate control; contractual provisions for monitoring and bonding; conflicts of interest between creditors and stockholders; the distribution of knowledge within corporations; shareholder value; the structure of corporate bond, lease, and insurance contracts; and corporations' accounting policy choice.

†See "Positive Accounting Theory" in this book.

1. the assignment of decision rights (authority and responsibility) within the firm,
2. the structure of systems to evaluate the performance of both individuals and business units, and
3. the methods of rewarding individuals.

Today these three components—commonly referred to as the three legs of a stool—are familiar to every Simon student because they provide a systematic framework that can be applied consistently to analyze organizational problems and design more effective organizations.

In fact, a large part of the foundation on which the Simon School's approach to management education is based today can be traced to Meckling's economic approach to problem-solving, the famous agency paper, and the research it stimulated. Simon students, for example, are taught to view issues within firms in terms of self-interested behavior. Simon students also learn, as Meckling and Jensen pointed out in 1976, that there are conflicts of interest in any firm, but that the parties have an interest in resolving such conflicts in order for the firm to create more value.

The Context

In order to fully comprehend the impact of the original agency theory paper, however, one must understand what the major disciplines in business looked like prior to the paper's publication in 1976. Simon Professor James A. Brickley, whose own research on franchise contracts and other areas has been influenced by Meckling and Jensen, points out that finance, accounting, and organizational theory "were each in a much different state than they are now in terms of intellectual discipline."

He notes, "Accounting was not much of an academic discipline prior to the 1970s. It was descriptive—how to do debits, credits, etc. In terms of organizations, primarily in the management area, there was not very much rigorous analysis. Economists were focusing on prices and how markets operate but they were not looking inside the firm. The firm was simply considered 'a black box.' It was viewed as a profit-making entity."

To explain where the University of Rochester began to play a significant role in shaping business, Brickley recommends turning to three key developments in business history:

- Nobel Prize-winner Ronald Coase wrote a seminal article in the 1930s in which he posed the key question, "Why do organizations exist?" His broad answer was that while conceptually you can do all economic transactions directly in the marketplace, many resource allocation decisions are made within firms and other organizations. Some economic transactions are conducted less expensively in the marketplace, while others are conducted less expensively within organizations. Coase's major point was that absent transaction costs, organizational form does not matter. Therefore, the choice of organizational form depends on transaction costs. In order to understand the choice, one must understand the nature of the associated transaction costs.
- There was a long gap between Coase's work and a second major development in corporate finance, which occurred in 1958. Economists Franco Modigliani and Merton Miller published a paper in *The American Economic Review** arguing that absent transaction costs, a firm's value is independent of its capital structure.
- Throughout the 1960s and into the 1970s many business scholars sought to define the different types of costs that might explain organizations and financial policies, such as capital structure. That was when the University of Rochester began making an impact, as the result of the agency paper published by Meckling and Jensen. Their argument, simply put, was that capital structure matters. They showed through their agency-cost argument, for example, the positive effect that debt can have on company value.

Lasting Impact

As a colleague of Meckling until the scholar/dean's retirement in 1983, Simon Professor Clifford Smith remembers him as "a colleague who

---

*"The Cost of Capital, Corporate Finance, and the Theory of Investment."

governed with a set of [the most] thoughtful, constructive, insightful, innovative ideas as anybody I've ever met in my life.

"He had a vision of the unifying role that economics could play across the business curriculum. But one of the problems Meckling had to overcome [in meeting Wallis's challenge to build a first-class business school] was to determine whether to just try and do good things and play by the rules of Harvard and Wharton or whether to do something different and distinctive and innovative. Bill said, 'We are too small to be all things to all people. We will not cover all the bases. We will do something with distinction and do it extraordinarily well.'"

One need only read the conclusion of the famous agency paper to realize that Meckling (who died in 1998 at the age of seventy-six) committed Rochester's business school to producing work that would have a lasting impact. The conclusion he and Jensen originally wrote in the first half of the 1970s could just as easily have been written at the dawn of the twenty-first century. It said, in part:

> The publicly held business corporation is an awesome social invention. Millions of individuals voluntarily entrust billions of dollars, francs, pesos, etc. of personal wealth to the care of managers on the basis of a complex set of contracting relationships which delineate the rights of the parties involved. The growth in the use of the corporate form as well as the growth in market value of established corporations suggests that at least up to the present, creditors and investors have by and large not been disappointed with the results, despite the agency costs inherent in the corporate form.
>
> Agency costs are as real as any other costs. The level of agency costs depends, among other things, on statutory and common law and human ingenuity in devising contracts. . . . Whatever its shortcomings, the corporation has thus far survived the market test against potential alternatives.

—Vicki Brown

13

Positive Accounting Theory

"It would have been a lot more difficult, if not impossible, to do what we did in a larger business school. We would not have had the interaction with colleagues in economics and finance."
—*Jerold L. Zimmerman*

Today's business leaders, managers, and scholars know that a firm's accounting methods, organization, and financial policy are as much a part of the technology a firm uses to produce products as are its production methods.

Yet few realize how the interactive and interdisciplinary nature of the graduate business school at the University of Rochester contributed to that understanding. Most are also not aware that the kinds of discussions business people have today about corporate financial policy have been enormously influenced by work done at the University. In fact, the role of accounting is vastly different and far more relevant to business success than it was widely perceived to be prior to the involvement of the University's business scholars beginning in the 1970s.*

Beginnings

The two teacher/scholars on the faculty of the William E. Simon Graduate School of Business Administration whose names are most closely

---

*A companion essay in this book, "Agency Theory," describes how the University's Simon School used an economics-based approach to make critical contributions in the areas of finance and organizational theory. The groundbreaking work discussed here was very much dependent on the agency theory research described in that essay.

linked with helping to move accounting to the level of a scientific discipline are Ross L. Watts and Jerold L. Zimmerman. Both professors on the faculty today, they arrived at the school in the early 1970s, at a time when accounting could be described as consisting of basically two camps.

The first camp included people who had been trained in traditional accounting programs by faculty worried about such questions as, "How should companies account for pensions?" and, "How should companies account for depreciated expenses?" It was the use of the word *should* that put this group in what was known as the "normatist" camp, because its members were seeking prescriptions (norms) that could be applied by accountants. The people who wrote normative literature were mostly concerned with making policy recommendations.

Watts explains that the large size of this camp was due primarily to the U.S. securities acts of 1933 and 1934. These acts established the Securities and Exchange Commission (SEC) and gave it responsibility for setting standards in accounting. "With little empirical support," Watts points out, Congress had ascribed the blame for the 1929 stock market crash to poor financial accounting practices, including a lack of uniformity across companies. The professional bodies to which the SEC delegated standard-setting for accounting "needed an intellectual framework for specifying accounting procedures."

Given the perceived deficiencies and the demand for prescriptions, many financial accounting academics thought a working theory of existing accounting was not as important as finding remedies for the deficiencies. The normatists assumed that accounting's major function was to provide information for stock and other capital markets. What was most important, in their view, was to disclose enough information for people to be able to determine what a company's cash flows looked like. To this camp, the choice of accounting method was largely irrelevant.

The Positive Accounting Camp

The second camp was much different from that of the normatists and came to be associated with the term "positive" accounting. The label was adopted from economics, and later finance, where it was used to distinguish research aimed at explanation and prediction from research directed at generating prescriptions.

This camp consisted of young faculty members, many of them educated at leading business schools. In the 1950s and 1960s a number of these schools had been the recipients of huge Ford Foundation grants, which had been awarded with the goal of improving the quality of research in business schools. By 1965, cutting-edge developments in finance at the University of Chicago and the growing availability of computers to test related hypotheses led to a dramatic concept: that capital markets (the marketplaces where financial assets are traded, such as the New York Stock Exchange) are *efficient*.

"Capital market efficiency," notes Zimmerman, "says that when information is publicly disclosed, the market will quickly impound it." For example, if two publicly held corporations announce they are merging, their share prices will quickly and accurately reflect that information. Thus, the window of opportunity for investors to use the information to make money is exceedingly small. Some would even say that "hot" news about a stock is worthless because it is already old news when you hear it.

By extension, with efficient capital markets, concerns such as how to time accounting charges or whether to reveal information in the body of a financial statement or in the footnotes are irrelevant. Accounting choice, per se, doesn't impact firm value. Zimmerman explains, "if a company buys an asset, how it depreciates it for financial reporting will not affect its underlying cash flows. While the depreciation method chosen for taxes will affect cash flows, firms can use different depreciation methods for tax and financial reporting."

Capital markets research ultimately became the catalyst for the demise of normative research. Yet, while the growing capital-markets research literature provided the positive accounting camp with important information on the relation between accounting numbers and valuations, it provided little explanation for and few predictions about management's choice of accounting methods. Empirical evidence suggested to Watts and Zimmerman, however, that choice did matter.

New Theory

"Ross and I asked ourselves, 'How does accounting matter to firms? Why would a company change depreciation for financial reporting but not for taxes? Why would they care, especially since evidence was accumulating that markets see through accounting changes? Why were the Big Eight accounting firms spending millions of dollars writing articles for the FASB [Financial Accounting Standards Board]? Why would rational people be spending these kinds of resources if accounting didn't matter?'"

Watts and Zimmerman's questions started them down a path that would lead to their first exposition of positive accounting theory in a 1978 paper, "Towards a Positive Theory of the Determination of Accounting Standards," in *Accounting Review.* Today their theory—enriched by years of work by scholars at the Simon School and institutions worldwide—is widely used to explain and predict accounting phenomena (e.g., choice of accounting methods, valuation of assets at cost or market value, and earnings management).

Broadly stated, the theory says accounting choices are not made in order to better measure financial components, such as earnings. Instead, accounting choices are made by individuals with specific business objectives and with a view to the impact accounting methods will have on the achievement of those objectives.

In 1979 Watts and Zimmerman won the American Institute of Certified Public Accountants Award for their groundbreaking first paper. The following year they won the same award for a second paper entitled, "The Demand for Supply of Accounting Theories: The Market for Excuses." A subsequent book that grew out of their research and the research of Rochester Ph.D. accounting students won the Kappa Psi Foundation Award for Distinguished Service and Achievement in Accounting in 1985.[1]

Today, an ever-growing body of positive accounting literature born from that original work provides corporate managers, public accountants, loan officers, investors, financial analysts, regulators, and others who must make decisions on accounting policy with a way to predict and explain the consequences of their decisions.

Developing the Theory

"What Ross and I were able to do," says Zimmerman, "is take advantage of our colleagues and the environment here [within the Simon School]."

Watts explains. "In almost all other schools, finance and accounting, marketing and other departments are separate and no one ever mixes, even in cases where there are tremendous gains to mixing. Here you have good economists and good quantitative people. That's a big advantage. You get to think more generally. And you see accounting as part of a bigger scheme of things—as part of bigger institutional arrangements."

Both Watts and Zimmerman point out that their positive accounting research was very much intertwined with the agency-theory research conducted by then Dean William H. Meckling and a Rochester faculty member, Michael C. Jensen.* Meckling and Jensen were already working on the concepts that would be embodied in their famous 1976 agency paper, "Theory of the Firm: Managerial Behavior, Agency Costs, and Ownership Structure," as Watts and Zimmerman struggled with the questions that would ultimately lead them to positive accounting theory.

Zimmerman says, "Up until [the publication of Meckling and Jensen's paper] economists tended to view firms as black boxes and assume firms maximized profits. That's fine for a Mom and Pop grocery, but if you have thousands of people they don't all show up and want to maximize profits naturally. You need incentives, supervisors, and a culture that makes it happen. Those kinds of incentives do not naturally just happen."

Through their discussions with Meckling and Jensen, the two University accounting scholars began to focus on the way individuals act in terms of their own self-interest—that is, making decisions by looking at the personal costs and benefits of an action.

Watts and Zimmerman also viewed accounting in the broader context of the firm. "We asked ourselves," Zimmerman says, "How is accounting used in the corporate governance process? If you view a firm as a collection of self-interested individuals, you require control mechanisms such as performance reports, incentive schemes, and asset-stewardship accounting. It turns out that accounting provides these controls."

Through empirical studies and extensive historical research, the pair determined that the three principal uses of accounting are in debt contracts (lending agreements), managers' compensation schemes, and the

---

*See "Agency Theory" in this book.

political process. The following statements, says Zimmerman, are examples of each of the three uses of accounting.

- Debt contracts—"If you go to the bank for a loan the bank will ask for an audited statement."
- Manager compensation—"Firms with managers who have had an exceptionally good year, well above their target bonus, will make it easier on next year by taking any write-offs this year. Today it is pretty much commonplace that professional managers have these kinds of incentives."
- Political process—"In the 1970s, when we were doing our research, you couldn't turn on the television without seeing stories of the alleged obscene profits oil companies were making. It made us think of accounting numbers being used in the media, and picked up by politicians, who start passing laws about which companies to target for antitrust, which gives firms incentives to choose mechanisms to reduce their accounting numbers."

Positive accounting research was influenced not only by Meckling and Jensen but also by many other members of the business school community. For example, work on bond covenants by Simon Professors Clifford Smith and Jerold B. Warner was very influential on the early empirical studies in which Watts and Zimmerman analyzed the impact of debt on accounting choice. "That's where Ross and I got our debt stories and understood the conflicts of interest debt holders have with shareholders," says Zimmerman.

Controversy

For many years, Watts and Zimmerman had to deal with considerable negative reaction while conducting their research, developing their initial theory, and making contributions to what would ultimately become a large body of positive accounting literature with wide acceptance around the world.

Ask Watts how receptive accounting experts were to the concepts that he and Zimmerman espoused, and he stops to tell a story that dates to the early days of their research efforts. The story describes the time Watts gave the keynote speech at a major accounting convention in his native Australia, prior to returning to that country for a fifteen-month teaching stint.

It was 1974 and I was thirty-one years old. I was on the podium with financial analysts and a few others and they liked the paper. But then there was silence. No one asked a question for quite a while. It was a long silence. Then the chairman of Coopers & Lybrand in Australia got up and boomed out, 'Of course, there's silence! They're all shocked that someone like Professor Watts is coming back to teach our students. To suggest that I run my accounting practice in self-interest is outrageous!'

The next day the Australian newspaper, the national paper, ran a headline in its financial section. The headline said, 'Professor shocks accounting profession.' The story went on to say that everyone in the audience, except for the waiters, were shocked that Watts was assuming all the parties in a firm look after themselves and accountants are no different than anyone else.

They were professionals, you see, who believed they acted in society's interest. It was a very difficult time for me for a while, but there were allies there. People were just not used to thinking like that.

Acceptance

One of the factors that played a huge role in winning wide acceptance for positive accounting was Dean Meckling's push for Watts and Zimmerman to launch a new accounting journal. Colleague Michael Jensen had started a new journal for Rochester, the *Journal of Financial Economics*, and it was very successful from the start. It also had a major, positive influence on how the world finance community viewed the University of Rochester business school. Meckling wanted the same kind of stature for accounting.

The *Journal of Accounting & Economics* (JAE) was started in 1978 and "clearly has had an impact," notes Watts. In its very first year of publication, it had a larger influence on the literature (as measured by the *Social Science Citation Index)* than either of the major accounting journals of the day. JAE, which is still edited by Watts and Zimmerman, has continued to outpace its competition ever since and has long been regarded as the nation's leading accounting journal and very influential internationally.

Watts notes, "When we first started we were publishing articles that were not being published elsewhere. It took a while for our competitors to publish the same kinds of [positive accounting research] papers. But now they all do."

Adds Zimmerman,

By calling it the *Journal of Accounting & Economics* we were specifically saying we view accounting as more than a narrow discipline in and of itself. That's because accounting is really part of an organization's fabric. Fundamentally it is a function of an organization, and key to the way the organization operates.

Look at the way the profession is going today. If you were a corporate controller twenty years ago your function was just to produce statements. Today accountants help produce reports managers can use to reduce costs. They have to find out what the major strategies of the organization are so they can provide information to people to help them do a better job. Accountants help create more jobs, better products, and more wealth for shareholders.

—Vicki Brown

14

Eradicating Childhood Bacterial Meningitis

"There were hundreds and hundreds of guys in labs working on antibiotic resistance. I thought, why not be one of the few to go after new vaccines?"

—*David Smith*

In 1987, the invasive bacterium *Haemophilus influenzae type b* (Hib) was the major cause of childhood bacterial meningitis. Over 20,000 children in the U.S. were infected and the rate was rising; 12,000 developed meningitis, 1000 died, and some 6000 others became deaf, blind, paralyzed, or mentally retarded.

Today the incidence of the Hib meningitis has been reduced by over ninety-eight percent. Bacterial meningitis was overwhelmingly a childhood disease, the median age of its victims fifteen months. Thus in one decade, a deadly disease has been virtually eradicated among all children in the U.S., Australia, and several European countries. The vaccine that accomplished this was the first one the FDA had approved for infants since the polio vaccine in 1955.

In 1996 David H. Smith and Porter Anderson won the Albert Lasker Clinical Medical Research Award, the most prestigious in the country, for their work in creating that vaccine and for bringing it all the way from the lab to commercial production. The story of these remarkable developments, and the great medical advances they have led to, has a particular association with Rochester, since it was at the University that Smith and Anderson completed the work they had begun together in Boston, and it was in Rochester that Smith founded the company that would bring it to fruition.

At the University of Rochester medical school in the 1950s, Smith learned pediatrics from Gilbert Forbes and W. A. Bradford, and—in a crucial "year-out" internship in Herbert Morgan's microbiology lab—began his lifelong interest in infectious diseases. In 1976 he returned to Rochester as chairman of the pediatrics department. His avowed mission was to establish the national scientific reputation of a department already well respected on its clinical side. To that end he quadrupled the research space, recruited a number of strong young faculty, and trained many second-generation leaders in pediatric infectious diseases. Among his recruits were his collaborators on the Hib vaccine, Porter Anderson and Richard Insel. (Anderson has since retired to Florida; Insel is director of the Strong Children's Research Center and Pediatric Immunology at the University's Medical Center.)

Preparation

After medical school, Smith rose from intern to chief resident in pediatrics at Children's Hospital in Boston. His mentor and model teacher there was Charles Janeway, an early researcher on human immunity and something of a visionary in his emphasis on preventive medicine. Smith remembers him standing at the bed of a child dying of meningitis, saying "This can be prevented. . . . It's wrong to wait until we have to try to treat these children. . . . One of you [medical students] should try to find that vaccine." He urged Smith to take up the challenge, insisting that if he put his mind to it he could finish the job in three years.

This was not a natural career choice. Since the advent of penicillin and sulfa drugs in the 1940s the interest in vaccines had gone into decline. Concern over some bad side effects had caused immunization rates to fall off, and drug companies hit by costly lawsuits were discouraged from testing the vaccines. Furthermore, vaccines did not offer suitably attractive profit margins. A three-dose inoculation per child did not compare with a lifetime of repeated antibiotic treatments. Though the polio vaccine proved the notable exception, in general, from the

1950s to the early 1980s, the medical community's romance with anti-biotics made funding for vaccine research almost impossible.*

But as early as the 1960s Smith already knew some troubling things about antibiotics. His interest in the epidemiology of bacterial drug resistance stretched back to his medical school "year out" in Herbert Morgan's lab, and grew into his first research career. Between 1965 and 1971, both at Children's and at Harvard on a postdoctoral fellowship in the bacteriology lab of Bernard Davis, Smith coauthored over twenty papers on antibiotic resistance factors. He acquired a worldwide repu-tation for his work in bacterial genetics, specifically on the disturbing phenomenon of the *horizontal* transmission of genetic material within the bacteria—meaning, the spread of resistance factors to multiple an-tibiotics or drugs.

There was also a more personal reason for Smith's choice. In Davis's lab at Harvard he had come to realize how little he liked to run with the pack. Labs across the country were buzzing with work on antibiotic resistance, but few researchers were going after new vaccines. At that child's bedside Janeway had said it would take only three years. To Smith, who thought of himself as a "sprinter," that was alluring. Yet twenty years later his colleagues would be in awe of his tenacity.

In 1965, at the age of thirty-four, Smith went back to Children's Hospital of Boston as head of the Division of Infectious Diseases. There he took up the challenge of the Hib vaccine, and recruited Porter Ander-son, whom he'd known from Davis's lab, to come and help. They began their probe of the Hib in 1968; their first joint publication came in 1972. From then until 1990, with Anderson or with others, Smith pub-lished over thirty papers on *Haemophilus influenzae type b*.

The Vaccine

The story of their success involves their overcoming several kinds of odds, only one of which was the lack of interest, and therefore research

---

*And indeed, since the introduction of the "wonder drug" sulfanilamide in 1937, certain kinds of bacterial meningitis had been treated with a high degree of success. One strain, *N. meningitis*, had seen fatalities reduced from between fifty and eighty percent to less than twenty percent in the epidemics during World War II.

funding, by the National Institutes of Health (NIH) and the medical community in general. A larger challenge was the science itself. The funding problem was solved when Janeway found money for them from the Boston Area Foundation. The science, simplified, was this:

In the 1940s, Hattie Alexander had inoculated patients with Hib infection with antibodies created from rabbits she'd injected with whole Hib bacteria. But serious side effects often resulted from this crude serum treatment. "We knew that vaccines of the future should be highly purified units of the two or three key chemical components of the bacteria responsible for a given disease," Smith said. Given the refinement, potency could be increased manyfold, and toxicity greatly reduced. Smith and Anderson set out on an imaginative search to purify the mucous coating of the bacteria and then use that to vaccinate children.[1]

The road to speedy success looked clear enough. When their tests failed to produce the expected immune response in animals, the two men took a leap of faith and inoculated themselves. Within a week or two, both their systems were producing the antibodies; after more tests, the vaccine proved to be a great (ninety percent) success—but, mysteriously enough, only for children over the age of eighteen months. This was unexpected. "We couldn't figure it out," Smith said. The problem was that, for unknown reasons, "the sugar-based vaccine was slipping through without triggering a response in the still-immature immune system." The baby's immune system could not respond to the saccharides (sugar molecules) that coated the Hib bacteria.

According to Richard Insel, Smith and Porter were "presented with an unexplored area in immunologic response. [They] saw this as a challenging intellectual problem: What's required to actuate the immune system to make this antibody?"

Anderson finally came up with the idea of employing an old "conjugate" technique used in animal tests since the 1930s, though the usual chemicals involved were thought toxic for humans. The idea was to chemically bond a safe protein with the Hib polysaccharide and trick the immune system into thinking it was getting a protein instead of a sugar.

"What Porter did," Smith recalled, "was develop a unique system." Having converted the heavy polysaccharide into an oligosaccharide with a small molecular weight, "he refined a highly purified protein as a carrier, a protein that had not been chemically treated in any way. And then he used a chemical bonding system that created a tight, natural bonding between the protein and the oligosaccharide that worked beautifully."

By this point Smith had brought his team to Rochester. Insel, the immunologist, and Anderson, the microbiologist, were the lab-detail men. Smith, now busy chairing pediatrics and building up the department, remained intensely and cheerfully interested in the Hib project, but he was first and foremost the pediatrician, with his eye on the "big picture" of disease-prevention. The collaboration remains a model of scientific partnership.

The result of the partnership was the first "conjugate" vaccine ever developed. The first humans to test it were the developers themselves; the first child to receive it was the daughter of their colleague, Keith Powell. It would eventually become the first FDA-approved vaccination for all *infants* since 1955. Its success, as Insel put it, "changed the face of pediatrics in hospitals."

Praxis Biologics

But it isn't enough to invent a solution in the lab; you must also have it clinically tested, produced, and marketed. To do any of these things you have to get FDA approval. And here again, the tide of the times was against Smith and Anderson, regardless of the beauty and potential of their discoveries.

By 1983, having built up the pediatrics department's national reputation, Smith took another leap of faith. Neither the University nor any of the major drug companies were interested in underwriting the Hib vaccine at that time. So Smith quit his job, mortgaged his house, borrowed $1.5 million in seed money, hired a handful of technicians, scientists, secretaries, and graduate students, and set up shop in the old St. Agnes girls' school on the University's River Campus. He named the company Praxis Biologics (*praxis*, Greek for action, or can-do, and also an abbreviation for *prophylaxis*, prevention, the byword of pediatric medicine). Smith, according to Insel, was officially CEO and chief science officer, but unofficially also cheerleader, coach, manager, money raiser, FDA-cajoler, and team owner.

Smith's strategy was to secure licensing for the original non-conjugate Hib vaccine for children over two in order to finance the development of the conjugate vaccine for infants. By 1985, the children's vaccine had been approved, and the first of the thirteen million doses were being manufactured in a windowless lab on the third floor of the Medical Center's O Wing. Shortly, production was moved to a plant outside Durham, N.C., which was later turned into a state-of-the-art facility.

Smith's strategy worked. The first vaccine was recommended for children two and older. This trumpeted success awakened the interest of the pharmaceutical industry and provided Praxis with the funds to pursue the target vaccine for infants. In 1990, after American Cyanamid bought Praxis for $232 million (keeping the company in Monroe County under the name of Lederle-Praxis), the FDA licensed development of the *Haemophilus influenzae b* conjugate vaccine for infants. The result was a dramatic, immediate reduction of Hib infections among children, and consequently of cases of meningitis.

The New Era

Thus the vaccines ushered in a new era of vaccine development. The changes were already in the air when Smith started up Praxis, knowing he would have to hurry to prevent the inevitable co-opting by one of the behemoth drug companies. The downside of the reliance on antibiotics was starting to show itself—the troubling increases in resistant strains of many diseases, the rising costs of health care, the attraction of the far cheaper route of preventive care. Explosive advances in immunology and recombinant DNA had also refocused interest on vaccine possibilities. But the dramatic success of the Hib vaccine in such a short time, and the very apparent promise of new conjugate vaccines on the horizon, changed the prevailing climate and cleared the way for research on a family of vaccines against venerable pediatric plagues like pneumonia, chicken pox, and ear infections.

By late 1999, the FDA had approved a new vaccine, based on the same conjugate principal used for Hib, for *Streptococcus pneumoniae* (another of the pathogens for meningitis) with a secondary effect against ear infection; and in England licensing was obtained for a vaccine against a killer strain of *N. meningitidis.*

The tremendously broadened horizon now includes the possibilities of vaccines for varieties of cancer, for type 1 diabetes, and even—further off but under active investigation—for the great killers tuberculosis, malaria, and HIV.

Rochester in particular has benefited from the boom in vaccine research that Smith and Anderson started. The University of Rochester in 1998 launched its new vaccine center, now aptly renamed the David H. Smith Center for Vaccine Biology and Immunology, with the express mission of integrating basic vaccine research with biotechnology development and public health programs. Here too the model is Smith's Praxis. "You can make the best vaccines in the world," says Peter Szilagyi of the Strong Children's Research Center, "but if you can't get them to the people who need them, they're worthless." Szilagyi himself created the program that has boosted the rate of immunizations at Rochester's inner-city schools.

For its part, the University helps pay for such expanded research facilities as the new Arthur Kornberg Medical Research Building through collaborations on the model of Praxis, from which it has so far received more than $30 million in royalties. The University furthermore is one of five sites designated by the NIH to evaluate new vaccines, which means the participation of local physicians and their patients as well as the research community as a whole.

In presenting the Lasker award to Smith and Anderson and to a pair of NIH scientists (John Robbins and Rachel Schneerson) who'd been working independently on the Hib vaccine, Joseph L. Goldstein said,

> The Hib vaccine has reduced the incidence of the Hib meningitis by 98% in less than 10 years—a truly remarkable achievement in the history of medical science. No other vaccine has ever shown such a rapid and dramatic effect in virtually eliminating a fatal disease. The story of our four awardees constitutes a wonderful chapter in the history of medicine—comparable to the eradication of smallpox by Edward Jenner, polio by Salk and Sabin, and mumps and measles by Maurice Hilleman.

In 1998, a year before he died of a rare melanoma, Smith predicted that within a decade whole chapters from the pediatric infectious disease textbooks would be erased. "The research that's been done during the last 20 years in the nation's laboratories—especially work that has opened the door to understanding the immune system—is revolutionizing the practice of medicine." We are on our way, he said, to seeing "the greatest impact on world health since the purification of drinking water."

—John Blanpied

15

Coping with the Stress of Illness

"I hoped to build a theory that would apply broadly to threatening healthcare events and to patients with different characteristics."

—Jean Johnson

The once-pervasive concept of the "omniscient physician"—proprietary guardian of a patient's healthcare—was shattered in the second half of the twentieth century. A growing "patients' right to know" movement mirrored the anti-authoritarian spirit of the 1960s, and the phrase "informed consent" became not only a new term in the moral and medical lexicography but one backed by formidable legal muscle. For most doctors, this kind of open exchange with patients was a new (and often uncomfortable) phenomenon. Generations of doctors worldwide had been conditioned to put into practice the old Latin adage that, at least where patients are involved, "too much knowledge is a dangerous thing."

During those contentious years of the 1960s, Jean Johnson, then a graduate student at the University of Wisconsin, began asking important new questions arising from her nursing practice: How were patients dealing with the information that now was being given them by their healthcare providers? What did they *really* need to know? What kinds of information would best help them cope with their illness and its treatment? When and how should that information be offered? How could nurses, who often deal more frequently and more intimately with patients than physicians, learn to become active and effective partners in and advocates for this new process?

Through an innovative series of research studies, conducted first in the laboratory and then in the clinical arena, Johnson created a model of information-sharing that has proven to be effective in minimizing patients' distress and enhancing their functional abilities as they undergo treatment. When your doctor or nurse carefully explains what you may expect to experience during and after the course of a medical or surgical procedure (the cold sensation of an anaesthetic salve, the noisy clatter of an MRI machine, the reddened skin that may accompany radiation treatment), you are seeing Jean Johnson's research findings translated into action. Much of our current understanding that patients must play an active role in their own health care derives from her research.

Johnson came to Rochester in 1979 from Wayne State University's College of Nursing, where she was professor and director of the Center for Health Research. As professor of nursing (now emerita) at the University of Rochester School of Nursing and director for nursing oncology at the University's Cancer Center, she has filled many roles: teacher, administrator, researcher, mentor. "Jean has turned the art of coping with patients' pain and stress into a science," says Robert J. Joynt, former dean of the medical school at the University of Rochester. "In addition, she has developed a corps of nurses who are essential to the outstanding patient care that is one of the hallmarks of our Cancer Center."

Self-Regulation Theory

For more than a quarter of a century, Johnson has been analyzing, testing, and promoting the best ways to help men, women, and children cope with the complexity of stress in the healthcare experience. Her research has had enormous clinical implications. She has documented that patients do best when they are prepared for their treatment with information describing what they will experience, information that is presented in concrete, objective terms. Conversely, she has shown that if patients focus on the emotional impact of an illness, concentrating on how painful the treatment will be and how it will upset their lives, they tend to expect these unpleasant feelings and interpret the experience in terms of the distress it provokes.

Johnson's self-regulation theory explains the process patients use to cope with physical illness. At the heart of the theory are four important assumptions:[1]

1. Patients want to know what to expect during a healthcare experience; they want to understand what is happening to them. Only then can they participate actively in their healthcare, regulating their own behavior in positive ways, rather than relying on others to direct their behavior.
2. Patients are motivated to achieve goals they have set for themselves. They want to minimize any negative impact of the healthcare event on the quality of their lives.
3. Care providers can acknowledge their patients' negative feelings, such as distress or fear, while helping them concentrate instead on the objective aspects of the experience. When patients focus on their emotions, energy that could be better applied to dealing directly with the objective experience is channeled away.
4. Patients who attend to the objective aspects of challenges that accompany illness are more likely to successfully minimize the impact of the experience on their daily lives.

The self-regulation theory assumes that each patient's interpretation of an illness will determine how he or she will manage an illness; that interpretation also will determine how satisfied he or she is with the outcome. Care providers can influence the ways patients cope with the experience, but patients' own conclusions and decisions are what guide their behavior.

On a practical level, Johnson's research has given doctors and nurses tools to help people deal effectively with their illness and the impact of its treatment. She has helped develop guidelines for caregivers[2] to use in selecting kinds of information (from the overwhelming amount of relevant information available) that helps patients manage their stress and discomfort. These interventions are predicated on the caregiver's describing the impending event *as the patient will experience it*, using concrete, objective terms. Vague, possibly threatening notions are replaced in the patient's mind with certain knowledge of what is about to happen. Studies by Johnson and others have shown that patient outcomes improved when the care provided by staff nurses was guided by the theory.

Johnson's First Experiments

During the late 1960s, the country was in turmoil over the Vietnam War and a growing insistence on civil rights. Jean Johnson, then a predoctoral fellow at the University of Wisconsin, began investigating how patients were affected when they were given descriptions of the physical sensations they could expect during a threatening healthcare experience. Clinical wisdom at that time predicted that telling patients what they would experience would increase their distress.

Johnson's interpretation of psychological research led her to question that hypothesis. In direct opposition, she hypothesized that knowing what to expect would reduce distress. Because the clinical wisdom hypothesis could have been correct, the first studies were done in the laboratory rather than the clinic, and the threatening event was pain from an inflated blood-pressure cuff. As Johnson predicted, the study revealed that subjects forewarned about the physical sensations they could expect found the pain less distressing than subjects who had not been forewarned.

Here, indeed, was new, substantiated information that held enormous clinical potential. Patients, it seemed, *could* benefit by receiving objective preparatory information. Laboratory results were strong enough to encourage Johnson to extend her research into clinical settings.

Johnson's first clinical study was conducted in partnership with John Morrissey, the chief physician in the endoscopy clinic at the University of Wisconsin. Morrissey was looking for ways to reduce stress among patients facing gastroendoscopy, a procedure which involves threading a flexible tube through a patient's mouth, down the throat, and into the stomach. Once again, the information-sharing technique was effective, both in the original study and in replication of studies. Patients who received an objective description of the events involved in the examination and the sensations they would cause required less tranquilizing drug than patients who received only the usual clinical care.[3]

"I hoped to build a theory that would apply broadly to threatening healthcare events and to patients with different characteristics," Johnson recalls of those years. To that end she extended her investigations to two diverse groups of patients: children, both male and female, who were about to have casts removed, and young women anticipating a pelvic examination. Both sets of studies supported the findings that forewarning patients of sensations they could expect to experience during the procedure improved their cooperation and ameliorated their distress.

The next step in the research was to study healthcare experiences that lasted for several weeks, a period during which patients' experiences would change. Johnson studied functional outcomes and emotional outcomes among patients who were given objective descriptions of what they would experience before, during, and after surgery; this information was withheld from a control group. The informed group of patients had shorter hospital stays and earlier resumption of their usual activities. Better still, evidence suggested that the informed patients used their new coping mechanism (concentrating on concrete, objective goals, rather than emotions) to improve the way they functioned after they were discharged.

Advancing the Research in Rochester

Johnson's appointment to the University of Rochester' nursing faculty in 1983 and her role as director of nursing oncology at the Cancer Center gave her an important new venue for her research, as well as a nursing research team skilled in providing cancer care.

"I wanted to find if the results of the research with surgical patients could be replicated in the case of still more complex healthcare events, that is, with cancer patients whose treatment and recovery from side effects span weeks," Johnson recalls.

During the 1980s, three studies of cancer patients receiving radiation therapy (RT) were conducted by Johnson and her team. The patients were studied during their RT treatments, which took place five days a week for four to seven weeks, and for two months after treatment, when side effects of the radiation continued to be observed. The first step in each study was to meticulously document the patient's experience. For example, researchers periodically interviewed a group of patients undergoing RT, documenting symptoms, recording what factors exacerbated or relieved symptoms, and at what time of day or week side effects appeared or receded.

With this information in hand, an RT intervention was developed and tested. The intervention was divided into messages, consisting of descriptions of the experience in concrete, objective terms, delivered by audiotape recording before each phase of the experience. The first message described what patients would experience during planning for RT. The second message, delivered just before the first treatment, included information about the size and location of the treatment room, the sound of the treatment machine as it rotated, and the length of duration of the first treatment. A third message, delivered during the first week of treatment, presented information about the nature, timing, and pattern of the most frequently experienced side effects of RT. The final message, delivered during the last week of treatment, described changes in side effects following the completion of treatments. (Suggestions for ways to reduce or control side effects were, of course, given to all patients, exclusive of their roles in the study.)

Johnson and her team then assessed the effects of the intervention, using reliable scales to measure the amount of disruption RT caused in patients' usual activities, as well as their emotional response. Johnson's findings were consistent in all three studies: Patients who received the interventions had less disruption in their usual activities during and following treatment than did patients in control groups. The experimental groups demonstrated significantly less disruption throughout (including thirty-three percent less during the last week of treatment, when side effects are usually most intense) and for one month following treatment.

Finally, Johnson's team designed a study to address the usefulness of self-regulation theory for practicing nurses. Staff nurses in the radiation therapy department at Rochester's Cancer Center were taught to use the theory with their patients before, during, and after treatment. Patients who received the theory-based nursing care had thirty-one- to sixty-percent decreases in disruption in their usual activities at various times of measurement as compared to patients treated by these nurses *before* they had participated in the learning exercise.

As a direct result of Johnson's research, the content of nursing curricula across the country has changed, as has nursing practice. At the national level, Johnson has played a leadership role in shaping the nation's health care—as a member of the governing council of the Institute of Medicine of the National Academy of Sciences; as a permanent member of two review committees of the National Institutes of Health; and as former chair and longtime member of the American Nurses Association's Council for Nurse Researchers.

The impact of Johnson's work on the way healthcare is provided can hardly be overestimated. A study of citations in nursing literature published between 1981 and 1993 shows that she is the author of five of the eighteen most-cited articles; she is also the nurse-researcher most cited in non-nursing journals. One of her early papers,[4] reporting research she conducted when a graduate student, has garnered 188 citations to date and has helped bridge the distance between the fields of nursing and psychology. In all, she has more than a thousand journal citations to her credit—a sure proof of the continuing impact of her work.

Recently retired dean of the School of Nursing Sheila A. Ryan was one of the young nurses influenced by Johnson's research: "When I was a young nurse, the climate was such that we never told patients anything that might upset them or cause them anxiety—from their diagnosis to the nature of their treatment and procedure expectations. It was most frustrating to see the fear and uncertainty in patients' eyes and to pretend that they just 'shouldn't worry.' Then came Dr. Johnson's research, proving that informed patients have better healing outcomes."

In 1993, Johnson received an award for her outstanding contributions to nursing and psychology from the American Psychology Association. The citation speaks to the importance of her work: "For creating a body of scholarship that is both so elegant and so important that it has been embraced by all disciplines concerned with mind-body issues and health care, resulting in improved quality of life for patients everywhere."

—Nancy Bolger

16

Education as Social Practice

"The main thing about schools is that they are one of the few *public* spaces in which people are engaged with each other in the interactional work of making *meaning.*"

—*Philip Wexler*

Of all the University's schools (arts and sciences, medicine, music, nursing, business, engineering) the Margaret Warner Graduate School of Education and Human Development is the smallest. But in its forty-year lifespan it has developed a distinctive profile and sense of mission, and become well-positioned for the tumultuous contests, both academic and practical, facing the critical field of education in the coming century.

Mission

The Warner School profile is a blend of the theoretical and the empirical grounded in a philosophy of education as—and this it wants to emphasize—social practice. Innocent as it sounds, the last phrase represents a marked contrast not only to past behaviorist models of education, but to the dominant modern model of the school as economic production machine. All of the Warner divisions—Teaching and Curriculum, Counseling and Human Development, Leadership and Administration—emphasize the School's intellectual seriousness: the rigor of its degree programs, the interdisciplinary character of its academic purview, its insistence on understanding education through the language and methods of philosophy, political economy, psychology, literary theory, history, and sociology. Its goal is "to use the tools of critical thinking to improve education at all levels and for all learners." What's *distinctive* is its view of education as embedded not in "society" but in *particular* societies. Practicing education means to recognize and share in the sometimes chaotic and problematic daily realities of the social situations in which teaching and learning occur, while also trying to understand them profoundly. The bridge between university and profession, scholarship and classroom experience, theory and practice, is at every point supported by the conviction that the social sphere of education is where the deepest work of human development transpires.

History

Although the School was born in 1958, it didn't come of age until the mid-1970s, and it has only reached its current stage of young maturity within the last decade. In its original form as the College of Education its purpose was frankly to help sustain America's global strength by producing skilled leaders in teaching and administration. In the 1970s, reflecting a shift in focus to basic and interdisciplinary research, it became the Graduate School of Education and Human Development and added a Ph.D. program to complement the Ed.D. Under the deanships of Walter Garms (1980-1983) and Guilbert Hentschke (1983-1988), it began to reach inward, forming ties with the College of Arts and Science and the College of Engineering and Applied Science and with other professional schools, notably medicine and nursing, to encourage interest in improved teaching campus wide. At the same time, it reached outward to establish collaborative programs with the Rochester City Schools and other local educational institutions, launching its tradition of testing educational theory against day-to-day classroom realities.

The School's intellectual pioneer was William F. Pinar, who came to Rochester from Ohio State in the early 1970s. Trained in philosophy and literary studies, he was passionately critical of the traditional paradigm of education in the United States. This paradigm was, in his and other reformers' views, marked by a moribund allegiance to behaviorism, scientism, and regimentation, divorced from cultural reality, politically neutered, and empty of genuine content. In May 1973 Pinar, under the auspices of the College of Education, invited one hundred reform-minded scholars in the U.S. and Canada to the first Conference of Curriculum Reform, thereby launching the "Reconceptualization" movement. Annual conferences followed at other universities. In 1978 Pinar founded the national *Journal of Curriculum Theorizing* (JCT), effectively institutionalizing the movement.

As in any revolution, the movement's themes were vigorously contested from within, but in general they embraced the postmodern idea of curriculum as historical, political, and autobiographical text (as against the hierarchical, "scientific," "news-from-nowhere" assumptions of the traditional paradigm). The curriculum was also to be centered on the whole human being (aesthetic and rational as well as technical), and derived from theoretical disciplines of the humanities, psychology, social and political theory (as well as the cognitive sciences). One major effect of this "paradigm shift" in curricular theory—judged by Pinar to be essentially accomplished by 1981—was to bring vastly more scholarly pursuits and political perspectives, hitherto ignored, to bear upon educational matters.

In 1989 Philip Wexler succeeded to the deanship.* A political and social theorist, he had come to Rochester in 1979 on a joint appointment in education and sociology, bringing a background and beliefs in the social content and context of education, and a postmodern view of society, not as a machine or a structure, but as an arena of social interactions. His project, as dean, teacher, and writer, has been to add to Pinar's political and humanistic legacy a strong vein of social theory and an understanding of education as social practice.

(Another thread of scholarly research at the School, begun years ago but still strongly pursued today, relates to education and political economy—that is, the effect of public and economic policy, including legal issues, taxation, resource allocation, and cost effectiveness, on educational institutions and practices. Over the years this approach has been followed by Walter Garms, William Boyd, and, currently, Tyll van Geel and Brian O. Brent.)

---

*Wexler served as dean through June 30, 2000, when he took a leave from the Warner faculty to join the Shalom Hartman Institute, an interdisciplinary, advanced research center in Judaic studies located in Israel.

In 1993 the School was endowed by William F. Scandling and named for his wife, Margaret Warner, both of them strong advocates of the School's balance of social practice with a field-oriented professional approach. This marks the point when the School came into its still-youthful maturity. Such a major named endowment was a significant recognition of the School's seriousness of purpose. And of course it also allowed the Warner School to fund more solid research projects and attract more young, high-quality, future-oriented faculty.

Theory-Informed Research

The Warner School takes theory seriously, but insists that it be grounded in practical school-based experience. A notable example is the "Becoming Somebody" project which Wexler and several colleagues carried out in three Rochester area schools in the 1980s.[1] The team spent several hundred hours each interviewing students, teachers, and parents in a middle-class professionally oriented school, a mostly white working-class school, and a mostly black inner-city school.

"In my view," says Wexler, "the so-called 'education crisis' is really the leading edge of a much larger crisis of public and institutional life" including economic, family, religious, and political structures. It is "first and foremost a crisis of *public life* in the U.S. . . . The main thing about schools is that they are one of the few *public* spaces in which people are engaged with each other in the interactional work of making *meaning*. They are places for making the *core* meaning, of self or identity among young people."

He and his team discovered that the central activity in the schools was not the skills-acquisition of the economic-corporatist model, but "identity-work"—the work of "becoming somebody," forming an identity that has currency or value in the society of the particular school. "Within the intense social life of the school organization, some personal resources were ignored, while others were seized upon, used, and affirmed as collectively valuable—building up in the process their possessors' image of identity. Identity was the pay-off for deposit of organizationally usable interactional resources."

Although the nature of the identity work was different according to the different social organizations of the schools, in all cases, the authors concluded, it was carried out in the absence of a shared "public space." In the middle-class school, called Penbroke, "there is not simply an absence of identification with a social center; there is recognition of an absence, and a yearning to fill it. The students, like the teachers, are fragmented into areas of interest. . . . [All] surrender the social totality of the school."

In "Grummitt," the working-class school, *interaction* breaks down in a vicious circle. "'Nobody cares' is the result of a socially conditioned mutual withdrawal of emotion and identification by students and teachers."

> *Student:* It's a vicious circle, 'cause once the kids start bumming out on the teachers or the curriculum, or whatever, then they skip classes, or whatever, and then they have to come down harder on them, which makes the kids hate them more, which makes them come down harder and harder and pretty soon it gets out of control. And that's why two hundred people a day get called down for referral. . . .

> What [the teachers] describe is an historic moment of cutback and rationalization of teachers' work that unintentionally cuts out the central committed relation of mutual respect and caring between teacher and student. They describe how students see schooling and the pervasive "not caring." In doing so, they show also the erosion of the student-teacher relation from the teachers' side; a teachers' not-caring that they want desperately to resist, but seem unable to overcome. . . .

The interactional principle governing the withdrawal of commitment is the "likewise principle" expressed by this student:

> So, it's likewise, you know. The teacher doesn't want to teach the kids who don't want to learn. Kids don't want to learn because they don't like the teacher.

In Washington High, the highly stressed inner-city school, "students fight against the institutionalized process of emptying"—that is, the emptying-out of any coherent public space—"because their selves are openly at stake." ("I *am* somebody," is the defensive slogan here.) "Empty-

ing here is of the self, and it occurs less behind the backs of the students
than against their will, imposed forcibly and out of fear by their teachers and guards. . . . To the teachers for whom 'morale is in the pits,'
injuries to student selfhood are consequences of their need to manage
students, individually and collectively, that will rescue them from having to face the pedagogically onerous task of overcoming students' many
earlier deprivations or their own fears of students' violent uncontrollabil-
ity."

Becoming Somebody exemplifies what the institutional literature of
the Warner School now emphasizes, namely its "classroom-based" research. But the project was not simply field research or descriptive eth-
nography. Although not theory-driven, in the sense of being designed
along theoretical lines and then carried out to test them, it was defi-
nitely theory-*informed* in the signature tradition of the Warner School.
It was a project, you might say, one of whose major components was to
study and understand its own assumptions and methodologies.

School-Based Research

The Warner School faculty bring true interdisciplinary strength into
the arena of education theory and practice. There are joint appointments in history, philosophy, political science, and psychology, with
fields of expertise running from law and political economy, to educational history, social theory, language, and mathematics. Several faculty
members have direct experience in public schools as teachers, administrators, or counselors.

A review of selected research projects and collaborations illustrates
the range and variety of Warner School activities:

1. David Hursh, chair of the teacher education program, focuses on
the special nature and problems of urban schools. He has joined forces
with Monroe Community College (MCC) in a five-year Ford Founda-
tion-funded project to develop, implement, and evaluate ways to motivate inner-city students to go on to college. In a separate collaboration
with MCC, the State University College at Geneseo, and the Rochester
City School District, he is also exploring ways to attract new teachers to
careers in urban schools. "We need to change the way we think and talk

about urban schools," Hursh says. "Instead of thinking about urban schools as inferior to suburban schools, we need to think about them as places in which students bring different experiences that can be built on for learning." That recognition of schools as fields of *distinctive* social organizations (rather than as variations on an ideal model) is central to the Warner philosophy.

2. Howard Kirschenbaum, in the meantime, works on increasing parental involvement in the schools. Most agree that such involvement is beneficial, but *how* to achieve it is not so obvious in a climate of increasing demands for student performance and staff professionalization. Kirschenbaum, the Frontier Professor of School, Family, and Community Relations, joins forces with the Rochester City School District to implement the approaches emphasized by Partnership 2000, a national program now used in over nine hundred schools. By 2001, every city school will have in place the parent involvement team and parent-school liaison that Kirschenbaum and his team are recommending.

3. Meanwhile, Dale Dannefer, who has written extensively on theories of aging and human development, leads a three-year research project to study the results of major architectural, programmatic, and cultural changes in the Fairport Baptist Homes and the Jewish Home of Rochester. The changes were all aimed at improving the mental health and well-being of both residents and staff. The underlying premise is that nursing homes are also educational sites, in the broad sense of the term, which should be socially organized as communities of learning and living for whole human beings. "This project is part of an effort to hold society's feet to the fire about taking the notion of lifelong education and growth seriously," Dannefer says.

4. With funding from New York State, Joanne Larson, chair of the teaching and curriculum program, focuses her research on classroom language and literacy practices and examines the ways in which social and linguistic practices mediate literacy learning. With Spencer Foundation funding, she also has made an ethnographic study of the interaction of race and literacy in two elementary classrooms, one urban, one suburban.

5. Ivor Goodson, an expert in educational curriculum and reform, also works from a major long-term Spencer Foundation grant to explore the cultural, social, and political influences that bring change to

secondary school systems. Aided by Warner School doctoral students, Goodson (and a Canadian colleague) will examine six schools in urban and suburban locations in Ontario, Canada, and Rochester and interview teachers about their experiences in the 1970s and 1980s. The goal is to see how, and how deeply, newly emerging reforms and more diffuse social changes impact the cultures of the schools.

6. For Lucia French, the Third Church Head Start Program in Rochester is an on-going laboratory. Her innovative Preschool Curriculum for the twenty-first century, unlike the familiar literacy-based models of early childhood education, uses science as a key to school readiness. "It is an integrated curriculum with science at its core," that meanwhile focuses on developing social skills, attention-management skills, communication and listening skills, and problem-solving skills. "Science meets every child's need to understand the world through active investigation," French says. Warner School students analyze the language development and attention management of the children, while teachers apply the research and methods developed by French and others in their classrooms. "Lucia provides the theoretical background," says one teacher, "and we provide the day-to-day nitty-gritty."

7. The mounting pressure for standardized methods and assessment strains the innovative, flexible, human-centered, theory-and-research-based approach to education promoted by the Warner School. But such challenges, as Wexler sees them, provide opportunities to rethink routines. A notable example is the work of Raffaella Borasi, a mathematician, who has won five National Science Foundation (NSF) grants in eleven years to study and promote ways of improving mathematics instruction. She is currently working on a multi-year NSF grant in four Rochester-area middle schools. The project ("Making Mathematics Reform a Reality in Middle Schools") could hardly be more relevant, given the national anxiety about math and science literacy in the public schools. With several Warner colleagues, Borasi's $450,000 project is out to show dramatic improvement in the ways math is taught and students learn in the four schools, and ultimately to support nationwide reform efforts in math education.

On its face, mathematics would appear to be one part of the curriculum *not* addressable as "social practice" or "social knowledge." But, says Borasi, "you can't promote reform in the teaching of mathematics without understanding the forces playing out in schools." She has written extensively on the underlying theories of her approach,[2] challenging the traditional paradigm of the "transmission" of mathematics knowledge ("hierarchically organized, context-free, value-free . . . to be passed along by experts to novices") on humanistic, philosophical, and psychological grounds.*

Her inquiry-based approach, which she teaches to teachers as well as to students in the modeling program setting, looks at math as a science rather than as drill and practice. "It takes a different mindset," Borasi explains, "because you have to convince people that math is not just computation" but is, in fact, continuous with other forms of knowledge that is learned and used in interactional social situations.

These examples suggest the varieties of ways that the Warner School's social theory of education plays out in real-school settings. They also point to the cultural tumult in which American education necessarily transpires. And that is, Wexler would say, because schools are the quintessential public places of society. "What we see in these schools is . . . a problem in the institutional core of the public sphere, an erosion of the institutional mechanisms and processes that build social commitment. What is required is to rebuild the institutional core."[3]

For American educators to do that job while balancing the intensifying demands for both academic excellence and professional performance is a tall order, but one that the Warner School seems eager to take on. Though less renowned than the other leading schools of education such as Harvard, Minnesota, and Ohio State, the School is gaining a reputation among insiders as a place of high intellectual quality and innovative practice. And it is animated by its philosophic heart, its commitment to education as live social practice. If past is prologue, the Warner School is well-positioned for the future.

—John Blanpied

---

*"It is obvious . . . that attempts at changing the way math classes are currently taught and at introducing alternative pedagogies . . . are not likely to succeed unless we can provide a cogent critique of these assumptions and offer alternative views of knowledge, learning and teaching that are grounded in theory and research." ("Rationale for an inquiry approach to mathematics instruction," *Reconceiving Mathematics Instruction: A Focus on Error,* Norwood, NJ: Ablex, 1996).

17

Quantum Optics at Rochester

"The quantum state reflects not only what we know about the system but what is in principle knowable."

—Leonard Mandel[1]

Beginnings

Compared to other innovative ventures launched by President Rush Rhees—the Eastman School of Music in 1921, the School of Medicine and Dentistry in 1925, the opening of the River Campus in 1930—the birth of the Institute of Applied Optics* in 1929 was a modest event. No one could have guessed that the future marriage of optics with quantum mechanics would lead to Rochester's emergence as a world-renowned presence in one of the most exciting branches of physical science, quantum optics.

In 1929, optics was still a fledgling science in the U.S., in essence a branch of astronomy, and there was only slight interest in its other practical applications. Seventy years later, quantum optics inspires both worldwide scientific enthusiasm and popular amazement. The 1997 Nobel Prize in Physics was awarded for work in this field, and a 1970 University of Rochester physics-math B.S. graduate, Steven Chu, was one of the recipients. The field has also caught the attention of network television and *The New York Times*, where it is not unusual to find reports on the role of optical techniques in quantum computing, or production of the coldest gaseous form of matter ever achieved, or (*Star*

*"Applied" would later be dropped to reflect the expanded scope of the Institute's mission.

Trek fans take note) a recent demonstration in two laboratories of the first evidence for teleportation.

From the beginning, the Institute of Optics—the "First New World Optical Center," proudly proclaimed in the headline announcing its founding—was an academic-industrial partnership. During World War I, Kodak, Bausch & Lomb, and other optics-related companies had become alarmed by shortages of the materials, technical skills, and basic knowledge necessary for their operations. In 1919, George Eastman and Edward Bausch convinced Rush Rhees that the proximity of major optical companies made Rochester the ideal place for a university-business alliance for optical training and research. It took a decade of planning, but their vision was ultimately realized. Kodak and Bausch & Lomb each put up ten thousand dollars for equipment, and agreed to contribute twenty thousand annually for operations over the next five years—a bargain kept even during the Depression.

Thanks to this unprecedented tripartite partnership, and to the astute directorship of Brian O'Brien (1938-53), the Institute played a significant role in the World War II effort. O'Brien was awarded the National Medal of Merit by President Truman in 1948 for "his outstanding services . . . at the University of Rochester's Institute of Optics during World War II in developing . . . optical devices to reduce the losses of American forces in night warfare both on land and sea." In the postwar years, Institute scientists continued to make important contributions to modern optical technology. These included O'Brien's prescient insight in 1951 that *cladding*, in which a layer of reflective material is used to coat an optical fiber, could be used to decrease light leakage from the fiber,* and the development of computer-aided lens-design programs during the directorship of Robert Hopkins (1954-65). And during the 1970s, Michael Hercher and his Institute colleagues successfully converted the multi-chromatic dye laser invented by Kodak scientists into the accurately tunable device that has made modern quantum optics technically feasible. These and other accomplishments inspired President Robert Sproull to refer to the Institute of Optics as the "jewel in the crown" of the University.

*The first low-transmission-loss optical fibers would not be developed until the 1970s at Corning.

Optics in Transition

After World War II, the loose historical coalition between the Institute and the University's Department of Physics began gradually to reassemble itself into a new form.* Hopkins wanted to develop a future-oriented Institute with expanded research facilities and new appointments, among them researchers in areas of basic physics. In this, he was assisted by the complementary vision of Robert Marshak, who was expanding the focus of the Department of Physics beyond its historical emphasis on nuclear and particle physics. One of the first fruits of their cooperation was the appointment of David Dexter, who went on to conduct prize-winning research in solid-state physics and optics.

A joint appointment of even greater significance for the next forty years was made when Emil Wolf came to the Institute in 1959. A World War II refugee in England from Prague, he had studied math and physics in Bristol. Later in Edinburgh he became an assistant to the eminent Nobelist, Max Born, with whom he wrote the optical scientist's bible, *Principles of Optics* (1959). To Born's irritation, Wolf delayed publication in order to finish a chapter on the then rather esoteric theory of optical coherence. No one would be interested, Born growled. But the first laser was developed less than two years later, and Wolf's prescient addition became required reading around the world.

Giants of Modern Optics

A member of the physics faculty since 1961, but retaining his optics appointment, Wolf has provided the key continuous link between the two departments. But the arrival of Leonard Mandel as a visitor to the Institute in 1960, and his subsequent appointment to the physics faculty in 1964, was the crucial first step toward Rochester's prominence in the then newly emerging field of quantum optics.

---

*The Institute's home for forty years was in the Bausch & Lomb physics building, and for the first decade of its life, though a separate entity, it was run by the physics department under the directorship of physicist Russell Wilkins.

These two men—Emil Wolf, Wilson Professor of Optical Physics, and Leonard Mandel, DuBridge Professor of Physics and Optics—hold special places in the world of optics. They have been repeatedly honored internationally for their research, each having been awarded Frederic Ives Medals and Max Born Awards by the Optical Society of America and Marconi Medals by the Italian National Research Council. In addition, Wolf has received the Albert Michelson Medal from the Franklin Institute (U.S.), and Mandel the Thomas Young Medal from the Institute of Physics (U.K.). Between them, they have over five hundred research publications, including a landmark recent book, twenty-five years in the writing, *Optical Coherence and Quantum Optics* (1995), on the nature of light and the theories that underlie optical instruments, lasers, optical detectors, photon counting, and nonclassical states of light. They are co-organizers of the regular Rochester Conferences on Coherence and Quantum Optics, the premier such conference in the world. The first was held in 1960, and the eighth is being planned for 2001.

But while the scientific interests of Mandel and Wolf overlap in the general area of statistical optics, they are leaders of distinct branches of frontier research in optical physics. Wolf has been mostly interested in classical optics—the study of light as an electromagnetic wave. His unique contributions to optical coherence theory have brought him great renown, including seven honorary doctorate degrees from universities in seven different countries, and his work on the statistical theory of coherence has played a crucial role in the field of diffraction tomography (3-D imaging), a key technology in the development of the next generation of clinical imaging devices.

Wolf was the first deep student of what is known as "partial optical coherence," in which light is able to form diffraction patterns only indistinctly.* His most startling discovery in this area is known as the "Wolf effect," and describes how light from a distant source—such as a galaxy or quasar—may have its frequency "redshifted," or slowed, as the result of interference patterns caused by partially coherent light emission. The

---

*Optical coherence arises from the behavior of photon-emitting atoms. When they behave in unison as they do in lasers, their photons march in lockstep and they emit "coherent" light that diffracts crisply. In contrast, light is more typically produced by hot objects—e.g., light bulbs or the sun. In this case, the atoms behave randomly and independently and they emit only "incoherent" light.

implications of the Wolf effect could be profound, since redshifted light from galaxies and quasars is used by astronomers to calculate the speed and distance of remote objects, which are thereby observed to be receding at great speeds. Wolf says, "There's no question that the theory is correct. The question is, of what relevance is it to astronomy?"[2]

Quantum Optics at Rochester

Meanwhile, Leonard Mandel has been investigating nonclassical optics—in which light is described in terms of discrete particles known as photons—with equal success. His career, beginning in London and continuing at the University, has been directed toward one overarching goal, elucidating the properties and behavior of individual photons and collections of photons. Thanks in large part to the accomplishments of Mandel and his many Ph.D. students, what was once an esoteric topic now lies squarely in the mainstream of physics.

For four decades Mandel has designed and performed innovative experiments to understand the behavior of photons and how they interact with atoms. He and his students were the first to demonstrate the interference of single photons with themselves. Later Mandel, and his student Jeff Kimble, presently William L. Valentine Professor of Physics at Caltech, carried out the first experiments on photons from a nonclassical* light source. Most famously, Mandel has probed to the very heart of quantum mechanics with an experiment carried out in 1991, which put an interesting new twist on the Heisenberg Uncertainty Principle, the notorious dictum that to observe a quantum object is to change it.

In this experiment, Mandel designed what might be called a quantum pinball machine, a device which shoots photons at a beam-splitter and a series of crystals and mirrors. As a particle, the photon travels one of two distinct routes and therefore does not exhibit interference effects; as a wave, it travels both paths simultaneously and can thus interfere

---

*Photons emitted from a *classical* light source—e.g., any hot object—arrive correlated in time, a phenomenon known as *bunching*. Using a laser-excited two-level atom, Mandel and Kimble devised an inherently *nonclassical* light source in which photons are emitted uniformly, a property they predicted known as *antibunching*.

with itself. What Mandel showed was that it was not necessary to directly observe a photon's motion to cause it to switch from wave-like to particle-like behavior; the mere possibility of learning which path a photon will take forces it to behave like a particle and travel a unique route. Thus, says Mandel, "What matters in the quantum world is not what is known about a system, but what is knowable in principle, because this takes any anthropomorphism out of physics."[3]

The Future

Today, Mandel's and Wolf's Rochester optics colleagues are making unexpected, occasionally startling discoveries of their own—in the process, making the University of Rochester one of the world's premier institutions for the study of quantum optics. Joseph Eberly, Carnegie Professor of Physics and director of the Rochester Theory Center for Optical Science and Engineering, has discovered the limits of Einstein's Nobel Prize-winning photoelectric theory, which states that atoms emit electrons when bombarded with photons, and the more light, the more electrons. Using supercomputer calculations, Eberly predicted that atoms would begin to stabilize in superintense laser beams, making electrons less likely to be expelled. The first experiments, carried out in Europe, have tended to confirm this surprising prediction. For this and other contributions to theoretical optical physics, Eberly received the 1994 Charles Townes Award of the Optical Society of America.

Carlos Stroud, professor of physics and optics, hasn't been content to manipulate individual atoms. Instead, he has designed the means to exert unprecedented control of an individual electron inside an atom. In his lab, an electron was excited with short laser bursts to the limits of classical physics, where it behaved dualistically as both particle and wave. It remained in a well-defined orbit, but did so only by being in two well-separated positions at once. In this state the electron was induced to interfere with itself as if it were two different waves.

Ian Walmsley, professor of optics and recipient of the National Science Foundation's prestigious Presidential Young Investigator Award, also studies the boundaries between classical and quantum optics. He uses ultrashort light pulses to perform what has been referred to as "the equivalent of a CAT scan on a molecule." Under Walmsley's leadership, Rochester scientists have recently won a national competition for a substantial multiyear grant from the Department of Defense to establish the Rochester Center for Quantum Information, dedicated to the study of what has become the hottest topic in quantum optics today. The fields of computer and communications security and computing and information storage are widely predicted to be turned upside down by recent discoveries in quantum information science. For example, it is now known that the cryptographic principles protecting secure information channels, including those carrying the most secret military and diplomatic messages of the U.S. government, must be completely revised and the rules of current practice rewritten, in anticipation of the quantum optical devices that will in the future operate under nonclassical rules.

The newest member of the quantum optics faculty, physics professor Nicholas Bigelow, has quickly emerged as a world leader in research on the frigid fringe of the ultracold frontier. Winner of a rare "triple crown" of young investigator awards—a Sloan Fellowship, a National Science Foundation Presidential Young Investigator Award, and a Packard Fellowship—he brings individual atoms to a virtual standstill in his lab by imprisoning them with short laser pulses. This results in an atom puffed up nearly one thousand times beyond its normal size and supercooled to nearly absolute zero. "We have changed the very nature of the atom," Bigelow says. More recently Bigelow, taking the next step, has become the first to supercool an entire molecule close to absolute zero. What he intends with these experiments is to examine "atom optics," where wavelike atoms are used instead of photons for new imaging devices—microscopes, transistors, or even atom lasers. The "lenses" Bigelow designs for such devices seem a long way from those of the Institute's early years, yet they lie along the same continuum of innovative optical devices.

In 1951, Albert Einstein stated, "All these fifty years of conscious brooding have brought me no nearer to the answer to the question 'What are light quanta?' Nowadays every Tom, Dick, and Harry thinks he knows it, but he is mistaken." Nearly fifty years later, the quantum optics faculty at the University of Rochester have positioned themselves at the very heart of a remarkable worldwide effort that promises to finally answer Einstein's profound question.

—John Blanpied and Paul Slattery

18

The New World of Visual and Cultural Studies

"Visual Culture, once a foreigner in the academy, has gotten its green card and is here to stay."

—Susan Buck-Morss[1]

"Visual culture" can and does mean many things. At its broadest it implies a social transformation from a "typographical" culture in which the dominion of text, or the word, has given way to that of the image.

At the University of Rochester and elsewhere the concept of "visual and cultural studies" has moved into, and largely supplanted, the stately old discipline of "art history." The "visual studies" part of the rubric points in two directions. On the one hand it implies the expansion of the field of inquiry from a more or less established canon of fine arts production—painting, printmaking, drawing, sculpture, and lately photography—to include literally *all* the visual imagery of the modern world (but especially film and digital imagery). In the process, the field has freely appropriated the analytical tools of neighboring disciplines, notably literary theory, film studies, psychoanalysis, and anthropology.

But while the move to "visual studies" implies a hearty cross-disciplinary expansion of subject matter and methodologies, it also has entailed a decided reaction against the formerly dominant tradition of art history. The "new" historians applaud

a general tendency to move away from the history of art as a record of the creation of aesthetic masterpieces, which constitute the canon of artistic excellence in the West, toward a broader understanding of their cultural significance for the historical circumstances in which they were produced, as well as their potential meaning within the context of our own historical situation. . . .

The importance of the shift from the history of art to the history of images cannot be overestimated. . . . Once it is recognized that there is nothing intrinsic about [aesthetic] value, that it depends on what a culture brings to the work rather than on what the culture finds in it, then it becomes necessary to find other means for defining what is a part of art history and what is not.[2]

"The academy is visibly changing," wrote the editors of a ringing collection of essays in 1994. Indeed, what was an insurgent movement scarcely a decade ago has moved into the center of the field, feted and funded by foundations, government, and universities themselves. "We have become the establishment," wryly notes Michael Ann Holly, until 1999 the chair of the Department of Art and Art History and a principal architect of the Visual and Cultural Studies Program at the University of Rochester.

What, Why, How?

When it was launched in 1989, Visual and Cultural Studies (VCS) was the country's first such graduate-degree-granting program. Michael Ann Holly had come to the University in 1987 to chair and reorganize a fine arts department (as it was then known) in need of revitalization, and to develop a graduate program. But the genesis of the actual VCS program was somewhat serendipitous. There were no models for it, and its ultimate character depended in part upon the diverse interests of its multidisciplinary progenitors. Holly herself is a Renaissance-studies specialist who has written widely on critical theory and the historiography—the history of the history—of art. Norman Bryson and Mieke Bal came from comparative literature, Kaja Silverman and Constance Penley from English and film studies, Craig Owens from the art and art history department. Together the group happened to represent a synergy of interests in feminism, poststructuralism, psychoanalysis, historiography, film theory, and social-political issues of contemporary art. In addition, they all believed that visual studies, like literary studies in the 1970s and 1980s, was overdue for poststructural shaking-up. Holly, Bryson, and Keith Moxey, of Columbia University, organized a pair of National Endowment for the Humanities–funded seminars (at Hobart and William Smith College in 1987, and at Rochester in 1989) to sound the themes.

> Other humanities have not suffered as much as the history of art from institutional inertia. Literary studies, for example, has welcomed the unsettling. . . . The growing awareness of theoretical horizons shaped, for example, by class, ethnicity, nationality, sexual orientation, and gender, have compelled art historians to acknowledge our discipline's inseparability from a larger cultural and ideological world. . . . Within the academy itself, there has arisen a questioning of all the values it once safeguarded.[3]

So complete has the revolution been in scarcely more than a decade that it may be hard to appreciate how fluid and extemporaneous the VCS program continues to be. But as a truly interdisciplinary program, it depends for its particular strengths and offerings on the makeup of the participating departments. Of the founders, for instance, only Holly remained in 1999. Janet Wolff, a sociologist of culture from England, with an interest in aesthetics, arrived as director of the VCS program in 1991, and has played a major, shaping role in its development ever since. In the meantime they have been joined by, among others, David Rodowick, an English professor specializing in film; Lisa Cartwright, also from English and author of a book about medical imaging; Sharon Willis, who teaches French in the modern languages and cultures department (MLC) and writes about popular culture and film; Thomas DiPiero, also from MLC, a specialist in psychoanalysis, race, and popular culture; Robert Foster, from anthropology, working in global cultures and mass media; and Douglas Crimp, an art critic who also analyzes representations of AIDS.

Yet, Janet Wolff insists, through all its ad-hoc ebbs and flows and shifting dramatis personae, the program is bound by an internal vision that keeps it coherent and maintains its identity and its distinctive reputation. At its heart, she says, is a commitment to a vital feature of the contemporary academic world, namely critical theory.

What Is It?

What *is* the Visual and Cultural Studies Program at the University of
Rochester—technically "housed in" but in fact leading the transformation of the Department of Art and Art History? According to its own
literature, it

> provides students with an opportunity to study critically and analyze
> culture from a social-historical perspective . . . stresses the close interpre-
> tation of artistic production within historical and cultural frameworks . . .
> [and] offers students the chance to earn a master's or doctoral degree by
> doing intensive work simultaneously in several of Rochester's humani-
> ties departments.[4]

All students participate in the Visual and Cultural Studies Colloquium. Beyond that they take four courses each in visual studies and
critical theory, and six electives from courses in art history, film, and
comparative literature. The core courses in Visual Studies might include
(among about twenty possibilities) *Critical Theory in Art History, Representing AIDS*, and *Media Studies*. Courses in Critical Theory (among three
or four dozen) might be: *Feminist Cultural Studies of Science and Technology, Freud and Lacan (and Marx)*, and *Culture, Consumption, Consumerism*.

Clearly, the boundaries between "visual studies," "critical theory,"
and "cultural studies" are quite porous. In fact, "boundaries" themselves
are a subject of high feeling in the field, as in the debate over whether
"visual culture" and "cultural studies" are or could ever be called "disciplines." A "discipline" has historical connotations of hierarchy and politically protected fields of inquiry. Whatever else VCS is, it is interdisciplinary not only by necessity, but also in spirit.

The program is small—it accepts only a few new candidates a year—
but robust. The students tend to be older than their counterparts in
other departments. Many arrive with masters-level work from other
schools. Several have already published. Quite a few receive grants from
extra-University sources.

Some recent dissertation topics give an idea of the range of what is
meant by the "studies" in Visual and Cultural Studies:

- The construction of notions of "the child" across a range of visual and literary texts.
- An investigation of seventeenth-century French painting in relation to ideologies of gender in that period.
- The development of experimental documentary film in the areas of third-world and minority representation.
- Media pedagogy, institutional discipline, and cultural dissidence in the electronic media culture.
- Property, identity, and the aesthetic in eighteenth-century landscape gardens in England.

The VCS program has put its vivid mark on all aspects of the art and art history program. For example, at a time when more and more art departments in the country are splitting "studio" from "art history," it's just the opposite at Rochester. Nearly a third of the students come to the graduate program already equipped with master of fine arts degrees, some with impressive exhibition experience. This makes them ideal vehicles, as teaching assistants, for carrying the theoretical interests of the program into the studio arts courses. One no longer learns sculpture or computer art or photography in isolation from the spirit of critical inquiry that infuses the core theory and visual studies courses. Everything—including the traditional genre and period subjects—is to be understood in its political, social, ideological, and cultural context. Nor is there any pretense that the tenor of the critical inquiry is apolitical. "This is not an agenda-less operation," the proponents are unashamed to say. Some do make strong links between their radical social agendas and their academic interests, but the real point is a poststructuralist axiom: that *all* culture, *all* cultural products, and *all* cultural structures, such as race, class, and gender, are political in the most basic sense of the term. Critical theory merely brings the political dimension to the foreground.

Recognition

In the first ten years since its launching in 1989, the VCS program awarded thirteen Ph.D.'s (with several more due in 2000 and 2001).

The alumni have gone on to impressive positions—many in academic departments of art, art history, film studies, and communications, but others "in the world," including three in museum posts, even though the program offers no museum-studies course *per se*. Two graduates have already published their dissertations as books, and a third is under contract. Another has had his work selected for the prestigious 2000 Whitney Museum Biennial exhibition.

As for the faculty, among other honors of recent years, Michael Ann Holly and Janet Wolff have won Guggenheims, Douglas Crimp a Rockefeller Humanities Fellowship at Columbia University, and Wolff, Lisa Cartwright, and David Rodowick fellowships at the Society for the Humanities at Cornell. Perhaps the most striking testimony to the young program's reputation is that the Sterling and Francine Clark Art Institute in Williamstown, Mass.—one of the most traditional, and distinguished, art establishments in the country—hired Michael Ann Holly for a two-year (and possibly longer) stint as Head of Research.

Other tangible signs of VCS's success are the variety of imitative programs spawned in other universities in the U.S. and Canada (the latest being at the University of California at Irvine). There have also been some major grants. In March 2000 the department was awarded one of six American Art Dissertation Awards from the Luce Foundation, which will give the school $25,000 every five years to distribute among its most promising doctoral candidates at work on dissertations in American art.

Perhaps most impressive were two $350,000 Getty Foundation awards, in 1998 and 1999, for two Summer Institutes in Art History and Visual Studies. These intensive five-week-long programs on new trends in art history each drew twenty scholars from former eastern-bloc countries and ten more from countries around the world to Rochester. With sessions on "Historiography and Poststructuralism," "Feminist Studies and Queer Theory," and "Museum Studies and Postcolonial Theory," the 1999 Institute encouraged the scholars "to take up new tools of critical theory instead of old tools, such as iconology."[5]

Some of the papers from the institutes were published in an online electronic journal of visual studies, *In[ ]visible Culture*, initiated by VCS graduate students, and funded by the provost. In its first issue, among scholarly essays, a slideshow by Sasha Yungju Lee deftly exemplifies many of the movement's themes. By superimposing her own eyes on theirs, she "Asianizes" the photos of American female Hollywood icons— "Marilyn," "Doris," "Barbra," "Grace," "Elizabeth," and so on—thereby reversing cultural polarities, so to speak, and making wry points about mass culture, race, gender, and colonialism all at once.

Cultural Studies

"Visual culture" does not necessarily imply "visual studies." At the University of Rochester, *studies* is the crucial word because it means to think about, critique, and theorize about visual culture and its products. And by that logic, once you have *visual* studies you also have *cultural* studies, because it is axiomatic in postmodern theory that you cannot study visual phenomena in isolation from their cultural context. "What does Visual Culture study?" asks Michael Ann Holly. "Not objects, but subjects—subjects caught in congeries of cultural meanings."[6]

Not that cultural studies means the same thing to everyone, not even everyone in the University's Program in Visual and Cultural Studies. It has some roots in the British tradition of cultural studies, some in cultural and comparative anthropology. Certainly a major source is the work of French intellectuals like the philosophers Michel Foucault and Jacques Derrida and the psychoanalyst Jacques Lacan, principal intellects of the now fundamental notion that all cultural products, structures, and relations, including those of race, class, and gender, are "discourses" or "texts" that can be deconstructed (using analytic techniques borrowed from literary studies) by those savvy enough to be able to read them. But there's also a strong tradition in cultural studies importantly shaped by Marxism, feminism, social history, and postcolonial theory.

Some worry that the term "cultural studies" has been Americanized— that is, drained of its robust original content in such institutions as the Centre for Cultural Studies in Birmingham, England, in the 1970s and 1980s, appropriated by literary studies, and become "a kind of textualism—a set of ingenious, and perhaps politically-informed, new readings of texts, but readings that are ultimately ungrounded, arbitrary, and shallow." These critics decry the detachment of cultural studies from real-world social movements and "its increasingly professionalized and rarefied life in the academy."[7]

Still, the major themes of the union of visual and cultural studies are not in question:

- An image's cultural value is understood as contingent and formed in relation to its reception through time. Aesthetic value is not universal and transcendent, but rather meaningful only in a cultural context.
- The work is understood as actively organizing and structuring the social and cultural environment in which it was located, whether it is a Rembrandt, an English garden, or a psychosexual slasher fanzine.
- "Power relations . . . course through cultural representations." A work of art—or "image," or "representation"—can be "read" to show the way "class and gender work together to construct social difference."[8]

Amidst these general themes, however, an important debate is transpiring among those who practice visual and cultural studies. Janet Wolff's own project is to promote and define a "new cultural sociology" that can relate the aesthetic, the social, and the political in a kind of *post*-poststructural world. This is a world in which it is possible to revisit some of the old cultural ideas that had been swept away in the original revolutionary fervor. If the era of the imperiously autonomous masterpiece is over (along with the superbly detached spectator) it is still possible to wonder, for instance, about the "specificity" of the created object, about the ineffably enduring power that *some* objects have, and about the whole nagging subject of aesthetic pleasure. "It seems now," says Wolff, that "a somewhat over-enthusiastic leap into critical studies, and away from the object, is being remedied by a new attempt to combine critical thought . . . with respect for the object, acknowledgment of the *relative* autonomy of the aesthetic sphere and its practices and languages, and a commitment to careful readings."[9]

The Contentious Present

The revolution may be won, but meanwhile, inside the new domain, hot debate proceeds unabated. Arguments abound over terminology, over boundaries, over subject matter, over concepts like "privileged ocularity" and "the horizontal narrowness entailed in modernism's fetish of visuality."[10] To Hal Foster, "Cultural studies . . . sneaks in a loose, anthropological notion of culture, and a loose, psychoanalytic notion of the image." To Rosalind Krauss, "Students in art history graduate programs don't know how to read a work of art. They're getting visual studies instead—a lot of paranoid scenarios about what happens under patriarchy or under imperialism."[11] And these are visual-cultural *insiders*.

To Michael Ann Holly this contentiousness is just fine. Historians know as well as anyone the dangers of complacency. And besides, she and her colleagues have a taste for the "unsettled" they cannot conceal. In the journal *October*, she describes an eleventh-century sculpted tympanum at a church in Conques, France, depicting Christ sitting between congregations of saints on one hand, sinners on the other. The saints, orderly and rather complacent, contrast strongly with the crowd of sinners. "In a tumultuous cacophony of torment, figures of both sexes and all walks of life are beaten and eaten alive by rapacious monsters, plunged upside down into the flames of hell, and strung up by demons so fantastic that they all end up being much more engaging to dwell upon than their complacent counterparts on the other side."

Holly finds it too tempting to resist seeing the two crowds as depictions of "art history" and "visual culture" "as they stake out their semantic territory . . . at the end of the millennium."

> To display [a work of art] in a museum as a masterpiece, to locate it historically, or to seal its meaning shut by figuring out its iconography, is not to encourage it to continue to make meaning . . . in the present. In short, what we need to do is preserve the chaos of contemporary theory. . . . We need the disorderliness of spaces in conflict [rather than the neatness of linear "historic" time], the mayhem of the unknown, even if the resulting intellectual fracas sometimes seems like hell.[12]

Jonathan Culler describes two kinds of knowledge: the reproduction of the old, the production of the new. Visual and Cultural Studies is important, say its practitioners, because it challenges the uncritical reproduction of the old and promotes a new and more comprehensive kind of knowledge; and because it is an approach especially appropriate to the twenty-first century world and its visual culture.

—John Blanpied

19

Initiatives in Music for the Twenty-First Century

"For the day of the musician who knows music and music alone has passed, to be followed by the day when the musician must take his place in the world of men and affairs. A narrower training, no matter how excellent, no longer suffices."
—*Howard Hanson, Convocation, 1935*

"The biggest surprise of all is the discovery of how important music is to so many people, and how feeble have been our attempts, the attempts of the professional musician, to bridge the gap between our art form and the broader community."
—*James Undercofler, Inaugural Address, 1998*

"Who Killed Classical Music?" shouts the title of a recent book. It's how a lot of people in the classical-music world were feeling in the early 1990s. Audiences were dying off, funding drying up, CD sales falling, costs rising. Music programs in the public schools had been gutted. After several fat decades of growth, even great orchestras were struggling for funds and, perhaps more alarming, for signs of community interest. In the music schools and conservatories, professional musicians and educators had grown anxious about their futures. Students who'd been assured that "Practice! Practice! Practice!" would take them to Carnegie Hall were demoralized, worried, and confused.

At the Eastman School of Music a faculty commission, sharing the general sense of malaise, was set up to look into "outreach" possibilities and ways to attract new audiences. That was in 1993. Looking back from the year 2000, one would have to say that the climate for some much-needed change was favorable. Yet few would have predicted the speed of the revolution that has swept through the halls of George Eastman's stately school of music.

It is known as the Eastman Initiatives, and as with all revolutions, it raised early fears that costs might outweigh benefits—that musical excellence might be sacrificed to the god of audience outreach. The fears have not been realized. The Eastman School of Music, some eighty years after its founding, has emerged as a national leader not only in the continued production of top-flight musicians, teachers, and scholars, but also in its profound reshaping of the role of the professional musician in a rapidly changing world.

"Malaise" is very far from the spirit of the place these days. Exuberance is more like it, certainly as reflected in the two main pilots of the revolution, the new director of the School, James Undercofler, and the former dean of academic affairs, Douglas Dempster. Dempster came out of the humanities wing of the School, with a background in musicology, aesthetic theory, language, and philosophy. Undercofler, with a lengthy career as a musician and administrator, is passionate for music education. He was an Eastman student in the 1960s (in horn), went on to an M.M. from Yale and doctoral studies at the University of Connecticut, played first horn in the New Haven Symphony, founded and conducted the New Haven Concert Orchestra, and taught for and conducted the Greater New Haven Youth Orchestra. He is the founding director of the Minnesota Center for Arts Education, where he served from 1986 to 1995.

Undercofler arrived in 1995 as dean of academic affairs, specifically charged with implementing the proposals of the faculty commission convened by the former director Robert Freeman. In fact he did much more, becoming in effect the architect of the Initiatives. He immediately seized the opportunity to revitalize the battered old concept of music education, renovate an ossified curriculum, expand the canon of legitimate musical forms, and play some catch-up with new technology—this last both in the curriculum and in getting the school *wired*. (He had to hand-write memos when he arrived; now the entire School has internet access.)

Undercofler and his colleagues moved with extraordinary swiftness and boldness, mobilizing faculty and students to embrace the changes.

Already a number of workshops and extracurricular outreach activities had sprung up. Much harder was the next step—to nudge the de facto movement into the innards of the School's vital and rightly-guarded degree programs. This did meet some resistance—you do not make room inside an already packed curriculum without spilling some blood. Nevertheless, the Initiatives were launched with a clear set of objectives:

- to enhance the curriculum to meet the needs of today's musicians;
- to motivate and train students in audience-and-community outreach activities;
- to teach them to think as entrepreneurs about musical careers and train them for a much broader variety of professional positions than has been traditional;
- to challenge them to embrace with initiative and leadership the evolving styles of classical music and jazz;
- and, at the same time, to maintain the integrity of a program renowned for its commitment to the highest levels of instrumental virtuosity and musicianship.

The Initiatives

Music For All

Music For All was the original outreach Initiative and in some ways the most basic in its rediscovery of a nineteenth-century tradition of community-oriented performing. ("All music-making is local," as Dempster puts it—meaning that it exists and thrives within a real two-way relationship with its environment.) It is also probably the most comprehensive such program in the nation. All student chamber ensembles are required to give performances each semester out in the community—about a hundred a year altogether, mostly in schools, but also in malls, nursing homes, hospitals, community centers, and anywhere else they can reach nontraditional audiences. Their Eastman training prepares them with workshops and coaching on how to present their performances to these audiences. The benefits flow both ways: students learn to teach by doing, and they develop both performing and presentation skills by practical application, at no cost to the audiences.

The Ying Quartet—one of the most gifted string quartets in the country—is the model for this new relationship between artist and audience. The four Eastman graduates (and siblings), Janet, Phillip, David, and Timothy Ying, in the early 1990s fulfilled a two-year residency under the National Endowment for the Arts' Rural Residency Initiative, playing, teaching, and talking classical music in gymnasiums, malls, barns, and wherever else they could persuade the people of Jessup, Iowa, to listen. "A traditional conservatory education doesn't address the tools and skills we found we needed, as basic as learning to speak in public," says the violist, Phillip Ying. "How do you apply what you've learned in theory class to someone who has never even thought about a dominant chord?"[1] Now as faculty members of the Eastman School of Music, the quartet drives home the lesson to its students that audiences are as varied and complicated as the music they are hearing, and that musicians must be willing to give the time to educate them in whatever opportunity presents itself.

Another Eastman graduate, Catherine Denmead, spent an eight-week residency with her Florestan Quartet in rural southeastern Kentucky. "We had to think about what we wanted the children to take away from the experience. You are responsible for so much—you have to perform at a high level and have a *reason* to perform. You also have a chance to be responsible, to make an impact on someone's life."[2]

There are other ongoing outreach programs, such as "Music and Medicine: Partnerships in a New Key" (a series of small ensemble performances in hospital lobbies, oncology waiting rooms, and the like). All of them challenge the traditional notion that audience-building means only bringing people into the concert hall.

Curriculum review and reform

To Undercofler and others, it was clear that entering students needed to be taught musical skills beyond their specialties, and that those about to graduate needed, as Undercofler put it, "to be informed by those primary industries that comprise the classical music world." The question was how to make room for the new kinds of learning in a curriculum already bursting with the demands of the old kinds: basic musicianship, instrumental virtuosity and repertoire, analysis, history, and compositional comprehension.

Changes in the doctoral program came most easily, partly because they were relatively minor and you could quickly track the results. The basic innovation was to add a minor concentration—in early or sacred music, say, or in arts management—with the idea of widening the graduate's musical knowledge *and* career opportunities.

At the undergraduate level—by far the hardest nut to crack—the key was to redefine the core curriculum to allow more flexibility in the remaining space. Incoming students today are finer virtuosi than their counterparts twenty years ago, but narrower in their musical knowledge and range of skills. Therefore, music theory and music history have both been changed to reflect the Initiative ideas. Basic keyboard skills and chamber music components have been enhanced. A major new required course for freshmen, the Eastman Colloquium, gives the tightly focused students exposure to, and discussion of, the breadth of the School's departments. It also gives the faculty—not immune from the pressures of specialization—a chance to teach outside the regular curriculum.

The most far-reaching change for undergraduates will be the option at the end of their second year to choose either of two degree programs: the Applied Music degree (AMU), a performance-intensive major designed for students pursuing traditional performing careers; or the new Music Arts degree (MUA), an independent major designed from electives in the liberal arts, music history, music theory, courses in music and society, an individualized "minor" field of study, and a six-credit-hour senior project. The first opportunities to choose the MUA won't come until the spring of 2001, but early interest is running high.

Even before the new curriculum changes went into effect in 1999, departments were responding with their own initiatives. For their recitals, string students, for instance, now must play at least one piece written in the last forty years—by someone still living. Trumpet students must find people who have never attended a concert, help them choose one they are interested in, escort them to the performance, and then encourage them to attend another.

Double-degrees are also possible. A student might major in horn or strings or voice *and* English, math, or physics—acoustics, say.

Undercofler thinks the Initiatives program has already attracted better students than before—with higher talent, broader intelligence, greater knowledge of new kinds of music. And, he says, though technically advanced, they still play with "soul." A visiting composer was impressed with their "deep chops."

Arts Leadership Program

The Arts Leadership Program (ALP), founded and directed by Dempster, is a special curriculum to expand students' visions about careers in music and arts, to teach them survival skills in the rapidly changing world of professional music, and to train them to think beyond the narrow classical audition-and-performing career paths. Mini-courses in four general areas encourage them to take charge of their careers: 1) music and society, 2) arts administration, 3) presentation skills (knowing your audience), 4) the business of music (nuts and bolts). All ALP courses emphasize practical experience, and ask students to do some practical project—design a strategic plan for an orchestra, create a web page, conduct a social science survey, and so on. The courses are not required, except as preconditions for internships or the new MUA degree. After all, the essence of the Arts Leadership Program is to learn initiative.

The fact is, only twenty percent of even these elite students will go on to the kinds of performing and recording careers they may have come in dreaming about. The point is to give them some control over the business aspects of those careers, and to allow the other eighty percent the chance to arrive at other careers by their own choice, and not as a result of failure.

Students taking the ALP seminars will often go on to gain practical experience in senior-year internships, sometimes ones they have helped to design—in management offices of the Rochester Philharmonic Orchestra and other professional or semi-professional groups, and with government arts organizations, schools, hospitals, and such arts-in-education organizations as the Aesthetic Education Institute.

From there, they are excellently positioned for prized, highly competitive postgraduate internships, fellowships, or entry-level jobs. ALP graduates have taken residencies at Glimmerglass, the New York Philharmonic, and other prestigious institutions. Three recent graduates have won the highly competitive national fellowships of the American Symphony Orchestra League. Another went through the ALP directly into an internship with Chicago Symphony Orchestra; from there, though she was offered an arts administration post by the Chicago Symphony, she took instead a full-time arts liaison position with the St. Louis Symphony.

Eastman-city schools partnership

The partnership of the Eastman School with the Rochester City School District gives hands-on experience. It also helps the School fulfill George Eastman's original mission to enrich the community. In the process, it has begun to revitalize a once-renowned music education program gutted by decades of budget cuts and neglect. The partnership programs in recent years have included:

- "Time for Bows," the creation of Louis Bergonzi, an Eastman associate professor of music education. A rigorous screening process found twenty-six third-grade students from Rochester's poorest neighborhood and paired them with adult sponsors. Both student and sponsor committed to practice and play their stringed instrument for three to five years. In return, they received both group and private lessons from Eastman School faculty and graduate students.
- "Many Voices, Many Songs." Middle-school choral students were bused to the Eastman School and joined by Eastman students and faculty for choral training and practice in high-quality repertoire in an ideal setting.
- Community Education Division's Scholarship Program. Funded cooperatively by several city businesses and area foundations, this program gives up to seventy-five city students per year the opportunity to study tuition-free at the Eastman School's Community Education Division.
- The Ying Quartet's 1997-98 residency in School Number 17.
- The "Collegiate Pipeline Program." Eastman has committed to provide full-tuition scholarships to any Rochester City School District students who are admitted to the Eastman School, com-

plete a degree in music education, and are willing to return to the city school system to teach music. In return, the city schools have increased the number of music teachers and collaborated with Eastman in creating a number of special music programs.

Student-managed orchestras

Two notable examples of student initiatives over the last few years are Ossia and the New Eastman Symphony. Ossia, a fifty- to sixty-piece orchestra with an uncompromising contemporary-music repertoire, was created and is completely managed by a group of undergraduates with faculty advising. It launched in 1997 with an ambitious three-concert season, partly funded, thanks to James Undercofler, by the Institute of American Music, founded by the late Howard Hanson. For one of the works (György Ligeti's *Poème Symphonique,* involving one hundred metronomes) the group solicited the loan of metronomes in exchange for concert tickets. Metronome owners came, pointed out their babies onstage, and stayed for an evening of music most would never have expected to hear in their lives. After several successful seasons, Ossia has had to turn away students flocking to join the group.

The New Eastman Symphony (NES), created and managed by a ten-member graduate student group, mounts a full-scale performance schedule of traditional programs. Its mission statement declares, "If orchestras are to survive, the next generation of orchestral musicians must be equipped to rescue them." The NES thus exemplifies the trend whereby musicians, combating an often low morale, take responsibilities for their orchestras. "While the Symphony is an artistic success," says Douglas Dempster, " it is even more important as an experiment in student initiative and leadership."[3]

The Next Phase

In his inaugural address in 1998, Eastman's director James Undercofler didn't pause for self-congratulation, but focused on the "serious and widening gaps within the field that will eventually hinder a healthy relationship between music professionals and the public. . . . Exploring ways to close this gap will form the next phase of the Eastman Initiatives."

He identified two problems: one, the isolation and separation among music professionals; and two, the lack of a comprehensive, integrated vision that "includes music making, music learning, and music listening."

The isolation is already breaking down, thanks to the curricular changes and the other initiatives. Virtually every student and faculty member is involved musically in the community. The myth of the "world audience" for whom musicians perform is receding in favor of a richer interaction with real and specific audiences.

The "integrated model" of music education and music presentation is also under construction in the form of the various outreach programs, the Eastman-city schools partnership, and other working associations with the Hochstein School and arts-in-education programs like the Aesthetic Education Institute. The New Horizons Band, a part of Eastman's Community Education Division, allows over-fifties to take up an instrument newly and play in an Eastman-nurtured band. The Gateways Music Festival brings African American classical music and musicians to Rochester from around the world. All of these programs contribute to the aim of making "music flow naturally from classroom to concert hall, from concert hall to community music making."

It may take another ten years to confirm the nuts-and-bolts success of the Eastman Initiatives. In the meantime, there has been extraordinary national attention to the program, and despite its youth even some strongly traditional conservatory-style institutions appear to be considering its merits. There is the sense that in the truly brutal modern world of competition for funds and superior students, the Eastman School of Music has staked a genuine claim as one of the two or three premier music schools in the country, and is probably unique in its combination of traditional and contemporary strengths.

At the School itself, where the music-making is local, the atmosphere, says Undercofler, is "sensational," the sense of reinvigoration palpable. Students are accepting the "unknown . . . and the possible." Best of all, he says, the Eastman School has "re-energized the concept of one generation's breathing new life into the next."

—John Blanpied

Notes

Chapter 1: Inspiring America's Composers

1. Margaret Bond, "Howard Hanson Remembered," *Rochester Review* (Spring 1981): 3.
2. "New Director at Desk Here," *Rochester Democrat and Chronicle*, 7 August 1924.
3. Howard Hanson, "Modern Music and Its Problems: Second Article," *Rochester Democrat and Chronicle*, 23 November 1924.
4. Howard Hanson, "Eastman School Offers Composers Opportunity to Hear Own Works Played," *Rochester Democrat and Chronicle*, 15 January 1925.
5. "Works of Six U.S. Composers," *Rochester Democrat and Chronicle*, 11 April 1925.
6. Olin Downes, *Bulletin Celebrating the Fifth Anniversary of the American Composers' Concerts*, 1 May 1930.
7. Adelaide Hooker, "Rochester Sees 'Anthony Comstock'," *Modern Music* 11 (May-June 1934): 218.
8. Lawrence Gilman, "A Day of New Music, Domestic and Foreign," *New York Tribune*, 4 February 1924.

Chapter 2: Discovering Progesterone

1. George W. Corner, *Anatomist at Large: An Autobiography and Selected Essays* (New York, 1958), 61.
2. George W. Corner, *The Seven Ages of a Medical Scientist: An Autobiography* (Philadelphia, 1981), 119.
3. Ibid., 162.

4. Ibid., 231.

5. Ibid., 233.

6. Willard Allen's recollection, quoted by Corner, *Seven Ages*, 235.

7. George W. Corner and Willard M. Allen, "Physiology of the Corpus Luteum:II" and "Physiology of the Corpus Luteum:III," *American Journal of Physiology* 88 (1929): 326-46.

8. Corner, *Anatomist*, 49.

Chapter 3: Of Diet, Drugs, and the Cure for Anemia

1. George Hoyt Whipple, "Autobiographical Sketch," *Perspectives in Biology and Medicine* 2: 3 (Spring 1959): 288.

2. Ibid., 283.

3. G. H. Whipple and F. S. Robscheit-Robbins, "Blood Regeneration in Severe Anemia," *American Journal of Physiology* 72 (Feb. 1925). Authors' italics throughout.

4. Leon L. Miller, *George Hoyt Whipple, 1878-1976: A Biographical Memoir* (Washington, DC, 1995), 13-14.

Chapter 4: Breakthrough Chemistry:
The First Synthesis of Morphine

1. Dean Stanley Tarbell, *Autobiography* (Author, 1996), 71.

2. *Encyclopedia Britannica*

3. For my descriptions of the synthesizing process I am, in part, indebted to Marshall Gates, and for general information about the field to Jack Kampmeier of the Department of Chemistry at the University of Rochester.

Chapter 5: The Rochester Conferences on High Energy Physics

1. Quoted by J. B. French, "Robert E. Marshak: Tributes To His Memory," pamphlet (Rochester, 1993), 14.

2. Quoted in A. Pais, *Inward Bound* (Oxford, 1986), 19.

3. F. Szasz, *The Day the Sun Rose Twice* (University of New Mexico, 1984), 18.

4. R. E. Marshak, "The Rochester Conferences," *Bulletin of Atomic Scientists* 26 (1970): 93.

5. Proceedings of the Sixth Annual Rochester Conference (1956), viii-27.

6. Ibid.

7. J. C. Polkinghorne, *Rochester Roundabout: The Story of High Energy Physics* (New York, 1989), 57.

8. T. D. Lee and C. N. Yang, "Question of Parity Conservation in Weak Interactions," *Physical Review* 104 (1956): 254.

9. Polkinghorne, *Rochester Roundabout*, 71.

Chapter 6: The Biopsychosocial Model

1. George L. Engel, "The Need for a New Medical Model: A Challenge for Biomedicine," *Science* 196 (1977): 133.

2. Ibid., 132.

3. Theodore Brown, "The Historical and Conceptual Foundations of the Rochester Biopsychosocial Model," in Richard M. Frankel, Timothy E. Quill, and Susan McDaniel (eds.), *The Biopyschosocial Model: Past, Present, and Future* (University of Rochester Press, forthcoming).

4. John Romano, "Within Bareheaded Distance: The Story of Wing R, 1945-1975," in *To Each His Farthest Star: University of Rochester Medical Center, 1925-1975* (University of Rochester Medical Center, 1975), 300.

5. Ibid., 296-97.

6. Videotaped interview of George Engel by James Bartlett, M.D., 7 Jan. 1983, in *As Some Saw It: Interviews Recorded under the Sponsorship of the George W. Corner History of Medicine Society, University of Rochester, 1971-1992.*

7. Jules Cohen, M.D., "The Program in Psychosocial/Biopsychosocial Medicine," paper presented at conference on Health Professions, Education, and Relationship-Centered Care, San Francisco, May 1994.

8. Brown, "The Historical and Conceptual Foundations," 22.

9. Ibid., 23.

10. George Engel in nominating Robert Ader for an Albert Lasker Medical Research Award, 1996.

11. Robert Ader, "Research on Biobehavioral Aspects of the Model," in Frankel, *The Biopsychosocial Model.*

Chapter 7: Writers and Editors: Preserving the Canon

1. Emerson in His Journals," *The Journals and Miscellaneous Notebooks of Ralph Waldo Emerson*, vol. 1 (Cambridge: Harvard University Press, 1960), xxxiv.

2. *The Journals and Miscellaneous Notebooks of Ralph Waldo Emerson*, vol. 11 (1975), 519-20.

Chapter 8: Positive Political Theory

1. Riker's proposal for the new graduate program in political science as submitted to S. D. S. Spragg, William H. Riker Papers, Rush Rhees Library, University of Rochester.

2. "Department of Political Science 10-Year Report, Sept. 1973," William H. Riker Papers.

3. March 22, 1967, William H. Riker Papers.

Chapter 9: Geriatric Medicine's Coming of Age

1. Testimony on Preparation of Physicians for Geriatrics before the Subcommittee on Health and Environment of the Committee on Energy and Commerce of the House of Representatives, September 24, 1981.

2. University press release, May 11, 1973.

3. T. F. Williams et al., "Appropriate Placements of the Chronically Ill and Aged," *Journal of the American Medical Association* (10 Dec. 1973): 1333.

4. Kathy Quinn Thomas, "How Old Is Elderly," *Rochester Review* (Winter 1992-93): 12.

5. Robert L. Kahn and John W. Rowe, *Successful Aging* (Pantheon, 1998).

6. Thomas T. Perls and Margery Hutter Silver, *Living to 100: Lessons in Living to Your Maximum Potential at Any Age* (Basic Books, 1999).

7. Thomas T. Perls and John F. Lauerman, "Longevity's Opportunities," www.amazon.com (online reviews of *Living to 100).*

8. Ibid.

Chapter 10: The Unification Model of Nursing Education

1. Loretta C. Ford, "The Delivery of Primary Care," paper presented at the annual meeting of The New England Hospital Assembly, Boston, 27 March 1974.

2. "Unification Model of Nursing at the University of Rochester," *Nursing Administration Quarterly* (Fall 1981).

Chapter 11: Quest for Fusion: Laser Energetics

1. Mike Dickinson, "Fusing Scientific and Managerial Skills," *Rochester Business Journal* (8 May 1998): 10.

2. Moshe J. Lubin, in an interview with Joan Bromberg for The Laser History Project, 25 Sept. 1984.

3. Ibid., 7.

4. Malcolm W. Browne, "Reviving Quest to Tame Energy of Stars," *New York Times*, 8 June 1999, B-1.

5. R. L. McCrory, J. M. Soures, C. P. Verdon, et al., "Laser-driven Implosion of Thermonuclear Fuel to 20-40 g/cm3," *Nature* 335 (15 September 1988): 225-30.

Chapter 13: Positive Accounting Theory

1. *Positive Accounting Theory* (Prentice Hall, 1985).

Chapter 14: Eradicating Childhood Bacterial Meningitis

1. Nancy Bolger, "Vicissitudes of a Vaccine," *Rochester Medicine* (6 Oct. 1998): 13. Several of the following paragraphs are indebted to this interview-article as well as to an interview with Dr. Richard Insel.

Chapter 15: Coping with the Stress of Illness

1. Jean Johnson, *Self-Regulation Theory: Applying Theory to Your Practice* (Oncology Nursing Press, Inc., 1997).

2. Ibid.

3. J. E. Johnson, J. F. Morrissey, & H. Leventhal, "Psychological Preparation for an Endoscopic Examination," *Gastrointestinal Endoscopy* 19 (1973): 180-82.

4. Ibid.

Chapter 16: Education as Social Practice

1. Philip Wexler, with Warren Crichlow, June Kern, Rebecca Martusewicz, *Becoming Somebody: Toward a Social Psychology of School* (Falmer: London and Philadelphia, 1992). Quoted passages are from this edition.

2. E.g., *Learning Mathematics Through Inquiry* (Portsmouth, NH: Heinemann, 1992).

3. Wexler, *Becoming Somebody*, 156.

Chapter 17: Quantum Optics at Rochester

1. Quoted in John Horgan, "Quantum Philosophy," *Scientific American* (July 1992): 98.

2. Robert Kunzig, "The Wolf Effect," *Rochester Review* (Winter 1988-89): 18.

3. Mike May, "The Reality of Watching," *American Scientist* (July-August, 1998).

Chapter 18: The New World of Visual and Cultural Studies

1. Susan Buck-Morss, "Visual Culture Questionnaire," *October* 77 (Summer 1996): 30.

2. Norman Bryson, Michael Ann Holly, and Keith Moxey, *Visual Culture: Images and Interpretation* (Wesleyan Univ. Press, 1994), xv, xvi.

3. Ibid., xv. The same editors had also produced *Visual Theory: Painting and Interpretation* in 1991. Both derived from the NEH seminars.

4. The VCS Program's online description.

5. Scott Heller, "East Meets West in Art History: Scholars From Post-Soviet Countries Catch Up on American Trends," *Chronicle of Higher Education* (13 August 1999).

6. "Visual Culture Questionnaire," *October*: 40.

7. Janet Wolff, *"Cultural Studies and the Sociology of Culture,"* *In[ ]visible Culture: An Electronic Journal for Visual Studies (1999):* 7 (www.rochester.edu/in_visible_culture/issue.html*).*

8. Bryson et al., *Visual Culture*, xx.

9. Janet Wolff, *Aesthetics and the Sociology of Art,* 2nd ed. (Univ. of Michigan Press, 1993), 115. Italics added.

10. Thomas Crow, "Visual Culture Questionnaire," *October*: 37.

11. Quoted in Scott Heller, "What Are They Doing to Art History?" *Artnews* 96 (Jan. 1997): 105.

12. "Visual Culture Questionnaire," *October*: 41.

Chapter 19: Initiatives in Music for the Twenty-First Century

1 Robin Wilson, "To Help Its Students Find Jobs, Eastman School Expands Its Musical Repertoire beyond the Classical," *Chronicle of Higher Education* (14 March 1997), A-11.

2 Heidi Waleson, "MFA," *Symphony* (Sept.-Oct. 1997): 17.

3 Douglas Dempster, "The 'New Professionalism' in Music Education," Lecture at Keio University, Japan, 31 July 1999.

Contributors

S. M. AMADAE, a scholar on the history of science and technology, has written on the Rochester political science department and is currently working on a book discussing the history of rational choice theory in American political, economic, and policy science.

JOHN BLANPIED, a writer and editor in Rochester, New York, is author of several articles and a book on Shakespeare's history plays, *Time and the Artist*.

NANCY BOLGER is a freelance writer in Rochester, New York, and has been a senior medical writer and editor at the University of Rochester Medical Center.

VICKI BROWN is a Rochester, New York, freelance writer specializing in business, education, medicine, and the nonprofit sector.

ROBERT KRAUS is associate vice president for University public relations at the University of Rochester.

CHARLES E. PHELPS is provost of the University of Rochester.

FRANK SHUFFELTON is professor of English and American literature at the University of Rochester. His writing and research focus mainly on early American literature and culture.

PAUL SLATTERY is professor of physics and dean of research and graduate studies in the College at the University of Rochester. He conducts research in experimental high energy physics at the Fermilab accelerator facility (Batavia, Illinois) and served as chair of Rochester's Department of Physics and Astronomy from 1986 to 1998.